Training Circular
No. 31-29

TC 31-29
HEADQUARTERS
DEPARTMENT OF THE ARMY
Washington, DC, 9 September 1988

SPECIAL FORCES OPERATIONAL TECHNIQUES

TC 31-29

SPECIAL FORCES OPERATIONAL TECHNIQUES

September 1988

TC 31-29

Special Forces Operational Techniques

This Edition Copyright © 2013 by Special Operations Press

Cover, Layout and Design by: Special Operations Press

ISBN-13: 978-1481846516

ISBN-10: 1481846515

Proudly Printed in the
U.S.A

TABLE OF CONTENTS

ii

iv

7

v

8

PREFACE

This manual provides information and guidance on Special Forces (SF) tactics and techniques. It provides guidance for commanders, staff officers, and personnel responsible for SF training and operations.

It describes operational techniques used by SF organizations in the conduct of their missions in support of the AirLand battle. It also considers tactics that potential adversaries can employ against deployed SF units and methods these units can use to counter those adversaries.

Since SF soldiers must be thoroughly knowledgeable and proficient in conventional tactics and techniques, a large portion of this manual discusses or refers to conventional organizations, formations, and tactics. Special Forces use these tactics and techniques as a basis for instructing, advising, and assisting indigenous personnel in foreign internal defense (FID), unconventional warfare (UW), direct action, or SF reconnaissance operations. They also use them as a basis for developing detachment standing operating procedures (SOPs) and working documents.

The proponent for this publication is the US Army John F. Kennedy Special Warfare Center and School. Comments and recommendations for improvement to this manual are encouraged. Refer to a specific page and paragraph and include the rationale for any recommended changes. DA Form 2028 (Recommended Changes to Publications and Blank Forms) is to be used for this purpose and should be forwarded to: Commander, US Army John F. Kennedy Special Warfare Center and School, ATTN: ATSU-DT-PDA, Fort Bragg, North Carolina 28307-5000.

He, him, his, and *men,* when used in this publication, represent both the masculine and feminine genders, unless otherwise stated.

CHAPTER 1

SUPPORT OF MISSION REQUIREMENTS

Success in battle depends on the teamwork of confident, well-trained soldiers. This chapter prescribes the techniques for planning, preparing, and conducting combat operations in support of special operations (SO) in the AirLand battle to ensure that success. It is divided into two major sections: offensive and defensive operations. It is designed to be used by SF commanders and team members to support mission requirements.

Section I. Offensive Operations

This section describes those techniques the SF soldier must know to support mission requirements during offensive operations. The SF soldier must be able to plan and conduct raids, ambushes, and immediate action drills; establish patrol bases; track the enemy; and plan and conduct interdiction operations.

THE RAID

A raid is an operation, usually small-scale, involving a swift penetration of hostile territory. It ends with a planned withdrawal upon completion of the assigned mission.

Use of the Raid Force

It is used to—

- Secure information.
- Destroy or damage fixed installations or facilities.
- Destroy or capture weapons, ammunition, equipment, and supplies.
- Eliminate or capture enemy personnel.
- Liberate friendly personnel.
- Confuse, harass, and demoralize the enemy.
- Divert attention from other operations.
- Force the enemy to deploy additional units to protect rear areas.

Organization of the Raid Force

The size of the raid force depends on the mission, the nature and location of the target, and the enemy situation. The raid force may vary from a few personnel attacking a checkpoint to a battalion attacking a large supply depot. Regardless of size, the raid force consists of four basic elements: command, assault, security, and support.

Command element. This element is normally composed of the raid force commander and personnel providing general support to the raid, such as medical aidmen, radio operators, and, if a fire support element is part of the raid, a forward observer. The command element is not normally assigned specific duties with any element. It may be placed with any of the major elements of the raid force and wherever the raid force commander may best influence and control the action. When personnel who normally comprise a command element perform specific duties with an element, they are assigned to that element, and no separate command element is organized.

Assault element. The assault element is organized as determined by the mission and, specifically, by what is needed to accomplish the major objectives of the raid.

If the raid objective is to attack and render unusable a critical element of a target system, such as a bridge or tunnel, the raid force assaults and overcomes the key security functions. Immediately following the raid, the special task party places and detonates charges.

If the target is enemy personnel, the raid force may conduct its attack with a high proportion of automatic weapons, covered by supporting fire from the support element. In most instances the assault element moves physically on or into the target; in other instances it is able to accomplish its task from a distance.

Security element. The security element supports the raid by securing rallying points, giving early warning of enemy approach, blocking avenues of approach into the objective area, preventing enemy escape from the objective area, and acting as the rear guard for the raid force. The size of the security element depends on the enemy's capability to intervene and disrupt the operations.

As the assault element moves into position, the security element keeps the command group informed of all enemy activities, firing only if detected and on order from the command group.

Once the assault element begins its action, the security element prevents enemy entry into or escape from the objective area.

As the raid force withdraws, the security element conducts a rearguard action to disrupt and ambush enemy movement and pursuit and to create confusion by leading the enemy away from the main force's avenue of withdrawal.

Support element. The support element of the raid force may conduct diversionary or coordinated attacks at several points on the target to permit the assault element to gain access to the target. It also executes such complementary tasks as eliminating guards, breaching and removing obstacles to the objective, conducting diversionary or holding actions, assisting where necessary by providing fire support, and acting as demolition teams to set charges to neutralize, destroy, or render elements of the target unusable. Normally, the support element covers the withdrawal of the assault element from the immediate area of the objective, withdrawing itself on order.

Planning Considerations

Special Forces and resistance force commanders must consider the nature of the terrain and the combat efficiency of the raid force. Selection of the target is based on target criticality, accessibility, vulnerability, and recuperability (CARVER). The SF and resistance force commanders can assess the criticality and recuperability of various targets prior to infiltration during the area study and operational area intelligence study. Accessibility and vulnerability are situation dependent and these assessments must be supported by current intelligence.

1-2

Additionally, the raid force and resistance force commanders must consider any possible adverse effects on their units and the civilian populace as a result of the raid. The objective is to diminish the enemy's military potential. However, an improperly timed operation may provoke enemy counteraction for which resistance units and the populace are unprepared. Commanders must take every precaution to ensure that civilians are not needlessly subjected to harsh reprisals because of raid actions. An unsuccessful attack often may have disastrous effects on troop morale. Successful operations, on the other hand, raise morale and increase unit members' and their leaders' prestige in the eyes of the civilians, making them more willing to provide support. Detachment psychological and propaganda programs can exploit the impact of successful raids. Before such action is taken, however, any possible unfavorable repercussions from the population and the enemy military forces must be considered. If a raid is unsuccessful, psychological operations (PSYOP) personnel will be required to lessen any adverse effects on the friendly indigenous force.

Although detailed, the plan for a raid must be simple and not depend on too many contingencies for success. The raid force commander should plan activities so that the target installation is not alerted. Activities in the objective area must conform to normal patterns. The raid force commander should carefully consider times and space, allowing time for assembly and movement. He should consider all factors to determine whether movement and attack should be made during daylight or darkness. Darkness naturally favors surprise and normally is the best time when the operation is simple and physical arrangement of the installation is known. Early dawn or dusk is favored when knowledge of the installation is limited or other factors require tight control of the operation. A withdrawal late in the day or at night makes close pursuit by the enemy more difficult.

Medical support must be pre-programed, as reaction planning in the medical arena is predictably unsuccessful, unnecessarily resulting in loss of life or limb. Furthermore, adequate and visible medical planning has considerable positive psychological effect on the raid force's morale. Factors that should be considered include:

- Prescreening medical records and planning for special medical problems (for example, extra eyeglasses, medications as for blood pressure, immunizations).

- Pre-mission medical training for all personnel. A minimum of immediate first aid, as well as survival, evasion, resistance, and escape (SERE) medical considerations.

- Providing all personnel with self-care medical items, such as IV fluids, dressings.

- Planning treatment for anticipated nonbattle injuries or diseases incurred or contracted during the planning and rehearsal phases in the base area, as well as other assembly points.

- Planning to handle anticipated casualties of friendly forces, captured enemy personnel, liberated prisoners of war (PWs) at the objective, at planned rallying points, and in the base area. Considerations should include evacuation routes at all levels, priorities for evacuation, nonevacuation, and hospitalization, to include indigenous forces (if employed). Prior coordination with treatment facilities is necessary, but must be of a nature that does not divulge the nature or timing of the mission.

- Strategic placement of medical personnel within all elements (assault, security, support, as well as command elements) of the raid force.

1-3

Intelligence

The raid force commander must have maximum intelligence of the target, of the enemy forces capable of intervening, of the civilian population's attitude and support, and of the terrain to be crossed to and from the objective area. Therefore, an intensive intelligence effort precedes the raid. Resistance force intelligence and reconnaissance elements conduct pre-mission reconnaissance of the route to the target and of the target itself. In guerrilla operations, local auxiliaries may act as guides. The raid force begins surveillance of the target early and continues its surveillance up to the time of the attack. The raid force commander exercises extreme caution to ensure the secrecy of the impending operation. He carefully assigns missions to resistance force reconnaissance elements so that the local population will not become alerted and alarmed.

Participant Rehearsals

Using terrain similar to that found in the target area, when available, all participants conduct realistic rehearsals for the operation. Participants use sand tables, sketches, photographs, and target mockups to assist in briefings. They practice contingency and emergency actions and conduct final rehearsals under conditions of visibility expected in the objective area.

Final Inspection

The raid force commander conducts a final inspection of personnel and equipment before moving to the objective area. If possible, he ensures weapons are test-fired, faulty equipment is replaced, and the physical condition of each man is checked. He checks personal belongings to ensure that no incriminating documents are carried during the operation. This inspection assures the raid force commander that his unit is equipped and ready for operation.

Movement

The raid force plans and conducts movement to the objective area so that its approach to the target is undetected (Figure 1-1). Movement may be over single or multiple routes.

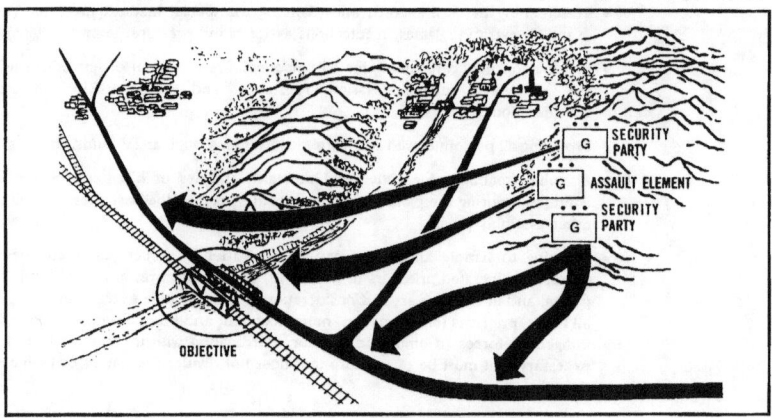

Figure 1-1. Movement to the objective area.

1-4

The preselected route or routes may terminate in or near one or more mission support sites (MSSs). The raid force makes every effort to avoid contact with the enemy during movement. Upon reaching the designated rendezvous and MSS, security parties deploy and make final coordination before moving to the attack position.

Action in the Objective Area

Special parties move to their positions and eliminate sentries, breach or remove obstacles, and execute other assigned tasks. The assault element quickly follows the special parties into the target area (Figure 1-2). Once the objective of the raid has been accomplished, the assault element and special parties withdraw, covered by the fire support elements with preselected fires. If the attack is unsuccessful, the raid force terminates action to prevent undue loss of personnel and the special parties withdraw according to plan. The assault and support elements assemble at one or more rallying points while the security elements cover the withdrawal according to plan. The assault element withdraws on signal or at a prearranged time.

Withdrawal

The raid force commander designs withdrawal to achieve maximum deception of the enemy and minimum danger to the raid force (Figure 1-3).

The various elements of the raid force withdraw on order, over predetermined routes to the base area through a series of rallying points.

Should the enemy organize a close pursuit of the assault element, the security element assists by fire and movement, distracting the enemy, and slowing it down.

If other elements of the raid force are closely pursued by the enemy, they do not attempt to reach the initial rallying point (IRP); but, on their own initiative, lead the enemy away from the remainder of the force and attempt to lose it by evasive action in difficult terrain.

Figure 1-2. Action in the objective area.

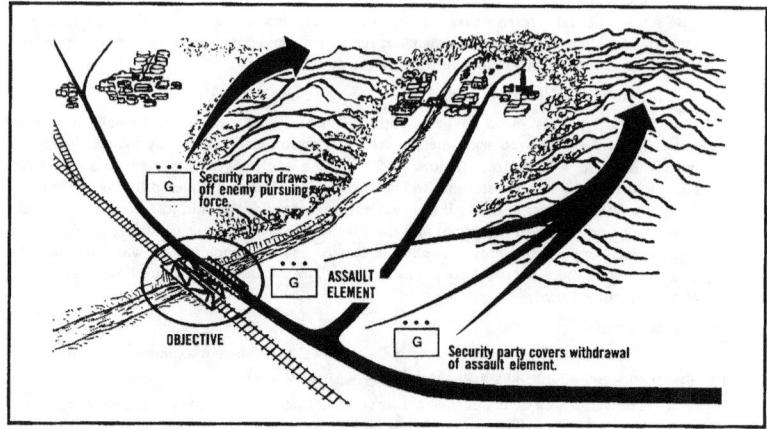

Figure 1-3. Withdrawal from the objective area.

The raid force commander issues specific instructions to the raiding force concerning such contingencies. Time–distance to be traveled, fire support and firepower, and the physical condition of the raid force personnel determine what course of action to follow. The raid force may attempt to reestablish contact with the main force at other rallying points, to continue to the base area as separate groups, to reach selected areas for evacuation or, as a last resort, to hold up in a selected MSS until such time as relief can be effected by the main force or a local auxiliary element.

The raid force, or elements of it, may separate and proceed as small groups or individuals to evade close pursuit.

Frequently in withdrawal operations, the raid force may disperse into smaller units, withdrawing in different directions and reassembling later at a predesignated place to conduct further operations. Elements of the raid force can conduct other operations, such as ambushing or pursuing the enemy force during the withdrawal.

Large Raids

When a target is large and well guarded, a larger raid force is required to ensure a successful attack. Large raids may involve the use of a battalion–size unit; and though conduct is similar to that of smaller raids, additional requirements must be considered.

Movement to the objective area. Surprise is just as desirable as in a smaller raid, but it is usually harder to achieve. In operational areas, the number of troops to be assembled and deployed may require additional MSSs at a greater distance from the target to preserve secrecy. In addition, a longer move to the attack position may be required. A large raid force usually moves by small components over multiple routes to the objective area (Figure 1–4).

Control. Control is an inherent problem in a large raid. Units without extensive radio communication equipment or units operating in an active electronic warfare (EW)

1-6

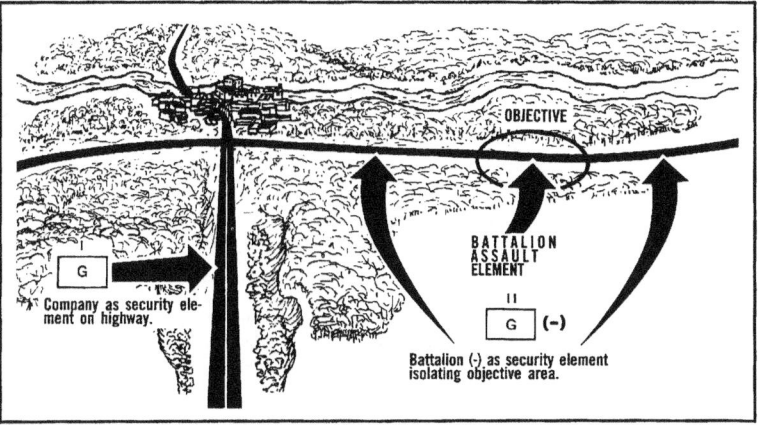

Figure 1–4. Movement to the objective area—large raid force.

environment will find coordination of widespread elements difficult to achieve. Pyrotechnics, audible signals, runners, or predesignated times may be required to coordinate action. In any event, even under optimum conditions, massing of the raid force at the objective is extremely difficult. During planning, the raid force commander must carefully consider the complexity of the plan and the possibility of overall failure if subordinate elements do not arrive on time.

Training. A high degree of training and discipline is required to execute a large raid. Extensive rehearsals help to prepare the force for the mission. In particular, commanders and staffs must learn to use large numbers of troops as a cohesive fighting force.

Fire support. Additional fire support is usually required. In the unconventional warfare operational area (UWOA) this may mean secretly caching ammunition in MSSs over a period of time before the raid. Each member of the raiding force may carry an extra mortar, recoilless rifle round, rocket, or box of machine gun ammunition, and leave them at the MSS or firing position for fire support units.

Timing. Timing is usually more difficult for a large raid. More time is required to move units, and the main action element needs more time to perform its mission. As a result, stronger security elements are required to isolate the objective for longer periods. The time the raid is conducted takes on increased importance because of the large number of personnel involved. Movement to the objective is usually accomplished during low visibility; however, because of fire support coordination requirements and the large number of personnel, the action may take place during daylight hours.

Withdrawal. In a UWOA, withdrawal from a large raid usually is done in small groups over multiple routes to deceive the enemy and dissipate its pursuit. Dispersed withdrawal also denies a lucrative target to enemy air and fire support elements. However, the raid force commander must consider the possibility of an alert and aggressive enemy

1–7

defeating the dispersed elements of the force. He must carefully weigh all factors before deciding how to conduct the withdrawal.

AMBUSH

An ambush is a surprise attack from a concealed position upon a moving or temporarily halted target. It is one of the oldest and most effective types of military operations. An ambush may include assault to close with and decisively engage the target, or the attack may be by fire only.

Ambush During Unconventional Warfare (UW) Operations

Ambush is highly effective in conventional operations. It is even more suitable and effective in guerrilla and counterguerrilla operations.

Ambush is a favorite tactic of guerrilla forces because it—

- Does not require that ground be seized and held.
- Enables small forces with limited weapons and equipment to harass or destroy larger, better armed forces.

Ambush is an effective counterguerrilla measure because it—

- Forces the guerrillas to engage in decisive combat at unfavorable times and places.
- Denies the guerrillas the freedom of movement on which their success so greatly depends.
- Deprives the guerrillas of weapons, ammunition, and equipment that is difficult to replace.
- Greatly weakens the guerrilla force through the death or capture of *hard core* personnel.

Purpose of the Ambush

An ambush is executed to reduce the enemy's overall combat effectiveness, to destroy, and to harass.

Destruction is the primary purpose of an ambush because a loss of men—killed or captured—and a loss of equipment and supplies—destroyed or captured—critically affects the enemy. The capture of equipment and supplies may assist our forces.

Harassment, a secondary purpose of the ambush, is also very important. Frequent ambushes force the enemy to divert men from other missions to guard convoys, troop movements, and carrying parties. When enemy patrols fail to accomplish their missions because they are ambushed, the enemy is deprived of the valuable contributions these patrols make to its combat effort. A series of successful ambushes causes the enemy to be less aggressive and more defensive minded. Thus, the enemy becomes apprehensive and overly cautious and reluctant to go on patrols, to move in convoys, or to move in small groups. It seeks to avoid night operations, is more subject to confusion and panic if ambushed, and in general, declines in effectiveness.

Types of Ambushes

There are three types of ambush: point, area, and hasty. For a detailed discussion of ambush formations see Appendix A. A *point ambush* is one where elements are deployed to support attack of a single killing zone. An *area ambush* is one where elements are deployed as multiple–related point ambushes. A *hasty ambush* is an immediate action drill (Appendix B).

Fundamentals of a Successful Ambush

Surprise, coordinated fires, and control are essential to a successful ambush.

Surprise. The ambush force must achieve surprise, otherwise the attack is not an ambush. It is surprise that distinguishes ambush from other forms of attack. Surprise allows the ambush force to seize control of a situation. If complete surprise cannot be achieved, it must be so nearly complete that the target is not aware of the ambush until it is too late for effective reaction. Surprise is achieved by careful planning, preparation, and execution. Targets are attacked when, where, and in a manner for which they are least prepared.

Coordinated fires. The ambush force must position all weapons, including mines and demolitions, and coordinate all fires, including artillery and mortars (if available), to achieve—

- Isolation of the killing zone to prevent escape or reinforcement.
- Surprise delivery of a large volume of highly concentrated fires into the killing zone. These fires must inflict maximum damage so that, when desired, the target can be speedily assaulted and completely destroyed.

Control. The ambush force must maintain close control during movement to, occupation of, and withdrawal from the ambush site. The ambush leader must effectively control all elements. Control is most critical when approaching the target. Control measures must provide for—

- Early warning of target approach.
- Withholding fire until the target has moved into the killing zone.
- Opening fire at the proper time.
- Initiating appropriate actions if the ambush is prematurely detected.
- Lifting or shifting supporting fires when the attack includes assault of the target.
- Timely and orderly withdrawal to an easily recognized rallying point.

Personnel conducting the ambush must control themselves so that the ambush is not compromised. They must remain still and quiet while waiting for the target to appear. They may have to forgo smoking, endure insect bites, thirst in silence, and resist sleeping, easing cramped muscles, and performing normal body functions. When the target approaches, they must not open fire before the signal is given.

IMMEDIATE ACTION DRILLS FOR FOOT PATROLS

A patrol may make contact with the enemy at any time. This is especially true in guerrilla operations. Contact may be by chance, by air observation or attack, or by ambush. It may be visual only; the patrol sights the enemy but is undetected by it. In this case, the patrol leader decides whether to make or avoid physical contact, basing his decision on the patrol's assigned mission and capability to successfully engage the enemy.

When a patrol's assigned mission prohibits physical contact, except that necessary to accomplish the mission, its actions are defensive in nature. It avoids being seen by the enemy. Physical contact, if unavoidable, is broken as quickly as possible and the patrol, if still capable, continues its mission.

1–9

When a patrol's assigned mission permits or requires it to seek or exploit opportunities for contact (as in the case of a search and attack patrol), its actions are offensive in nature and are immediate and positive.

In foot patrolling, especially in guerrilla operations, contacts (visual or physical) are often unexpected, at very close ranges, and short in duration. Effective fire, or the threat of effective fire, often gives leaders little or no time to fully estimate situations and issue orders. In these situations, immediate action drills provide a means for swiftly initiating positive offensive or defensive action, as appropriate.

Immediate action drills are drills that provide swift and positive small unit reaction to enemy visual or physical contact. They are simple courses of action in which all men are so well trained that minimum signals or commands are required to initiate action.

Immediate action drills—

- May be designed, developed, and used by any unit, no matter how it is organized.
- Are designed and developed as needed for the combat situation.
- May, in many cases, be initiated by any member of the patrol.

Immediate action drills are appropriate when—

- Ambushed, regardless of terrain.
- Contact, including an ambush, is at very close range and maneuver is restricted because of close terrain, such as mountains, jungle, or heavily wooded areas.
- Detection by air observation is a threat.
- Under low-level air attack.

It is not feasible to design an immediate action drill to cover every situation. It is better to know one immediate action drill for each of a limited number of situations that may occur in a combat area.

Some immediate action drills, such as those used in a counterambush, are initiated without signals or commands, as prearranged automatic reactions to enemy contact. Others, such as the chance contact immediate action drills, for example, immediate assault, are initiated on silent (arm-and-hand) signals. When appropriate, standard silent signals are used; *special silent signals are developed and used at other times.* There are no standard silent signals for *freeze, hasty ambush, and all clear.* The standard silent signals for *halt* and *enemy in sight (with direction indicated)* require exaggerated arm motions that increase the range of detection.

The special silent signals shown in Figure 1-5 may be used to halt a patrol in place, to indicate detection and direction of the enemy, to initiate the drills discussed here, and to indicate that all is clear.

Units that use other immediate action drills should devise and use special silent signals only when there are no appropriate standard signals.

Appendix B describes seven immediate action drills: one for any situation requiring an immediate halt, one for avoiding air observation, one for air attack, two for chance contact, and two for countering ambush. It is a guide to users in designing and developing immediate action drills appropriate to their situations. *NOTE: The drills and other actions*

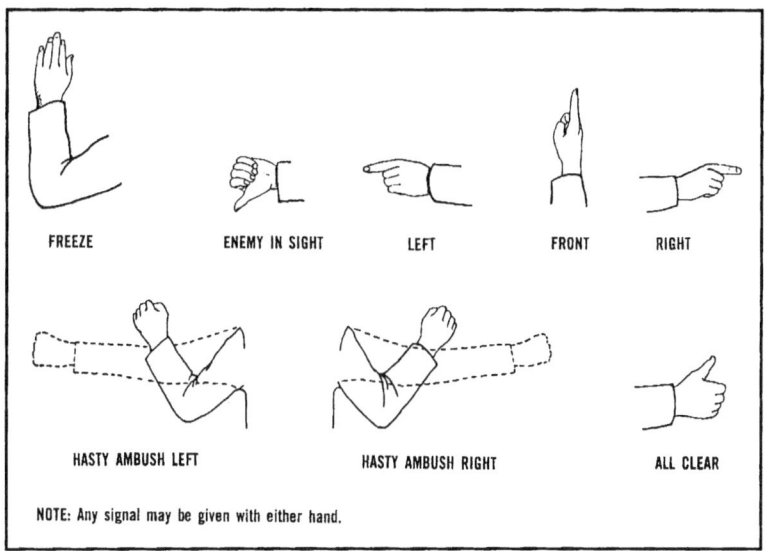

FREEZE ENEMY IN SIGHT LEFT FRONT RIGHT

HASTY AMBUSH LEFT HASTY AMBUSH RIGHT ALL CLEAR

NOTE: Any signal may be given with either hand.

Figure 1-5. Special silent signals.

described and discussed are examples of the application of principles. They are not to be considered standardized reactions that fit every situation.

Normal small unit tactics and techniques are used in executing immediate action drills and are not discussed in detail.

THE PATROL BASE

When a patrol is required to halt for an extended period, it must take active and passive measures to provide maximum security. The most effective means is to plan, select, occupy, and organize an area which, by its location and nature, provides passive security from enemy detection. An area so selected, occupied, and organized is termed a *patrol base*.

Planning a Patrol Base

Typical situations that require planning for the establishment of a patrol base include those situations where there is a need—

- To cease all movement to avoid detection.
- To hide the patrol during a lengthy, detailed reconnaissance of the objective area.
- To prepare food, maintain weapons and equipment, and rest after extended movement.

1-11

- To formulate a final plan and issue necessary orders prior to actions at the objective.

- For reorganization after a patrol has infiltrated the enemy area in small groups (used in conjunction with a rendezvous point).

- For a base from which to conduct several consecutive or concurrent operations, such as ambush, raid, reconnaissance, or surveillance patrols.

Any unforeseen situation occurring during a patrol could lead to an on-the-spot decision to establish a patrol base. When planning for a patrol base, the patrol leader must consider the mission and passive and active security measures.

Mission. A patrol base must be located in the best place for the patrol to accomplish its mission.

Passive security measures. The patrol leader should select terrain that is considered of little tactical value. The area should be remote from human habitation, but near a water source. Difficult terrain impedes foot movement. Dense vegetation, for example, bushes and trees that spread out close to the ground impedes foot movement.

The patrol leader should avoid known or suspected enemy positions, built-up areas, and ridgelines and topographic crests, except as necessary for maintaining adequate communications. He should also avoid roads, trails, and natural lines of drift and wet areas, steep slopes, and small valleys that may be lines of drift.

Active security measures. Plan for—

- Observation post (OP) and listening post systems covering avenues of approach into the area.

- Communications with outposts and listening posts.

- Defense of the patrol base, if required.

- Withdrawal, if required, to include multiple withdrawal routes.

- An alert plan.

- Enforcement of camouflage, noise, and light discipline.

- The conduct of necessary activities with minimum movement and noise.

Establishing a Patrol Base

The establishment of a patrol base is usually a part of an overall plan for a patrol's operation. In some circumstances, however, establishing a patrol base may be an on-the-spot decision.

The maximum length of time a patrol base may be occupied depends on the need for secrecy. In most situations, occupation should not exceed 24 hours except in an extreme emergency. *In all situations, a patrol base is occupied the minimum time necessary to accomplish its purpose.* The same patrol base is not usually used again.

In guerrilla operations, secrecy of the patrol base is mandatory. *A patrol base is evacuated if discovery is even suspected.*

In counterguerrilla operations, *secrecy* of the patrol base is always desirable, but is not always as essential as in guerrilla operations. The need for secrecy, and evacuation if discovered, depends on the degree of control the guerrilla force has of the area, its capability to react to the discovery of a patrol base, and its capability to adversely affect the patrol's operation. When the guerrilla force is relatively small and weak, secrecy of the patrol base may not be an essential counterguerrilla consideration, and discovery of the patrol base may not require evacuation. In an area controlled by a large guerrilla force with a relatively high degree of combat capability, secrecy of the patrol base is mandatory and evacuation is usually required, if discovered.

Selecting a Site

A patrol base site is usually selected by map reconnaissance during patrol planning. Selection may also be by aerial reconnaissance or may be based on prior knowledge of a suitable location.

A patrol base may be established on-the-spot as a result of reconnoitering, securing, expanding, and organizing the area occupied during a security halt.

A patrol base site, whether selected by map or aerial reconnaissance, or by prior knowledge of an area is *tentative*. Its suitability must be confirmed and it must be secured before occupation.

The patrol leader must also select an alternate site, a rendezvous point, and a rallying point.

Alternate site. The patrol uses the alternate site if the initial site proves unsuitable or if it must evacuate the initial site prematurely. During guerrilla operations, the patrol should maintain surveillance of the alternate site until it is either occupied or no longer needed. Reconnaissance and surveillance are not needed as much in counterguerrilla operations and are less likely to be possible or practical.

Rendezvous point. If the entire patrol evacuates the patrol base, it uses the rendezvous point. However, the rendezvous point will not have been reconnoitered.

Rallying point. If the patrol is dispersed from the patrol base, it uses the rallying point. This is the point over which the patrol has previously passed, has found suitable, and is known to all.

Occupying and Operating a Patrol Base

A patrol base may be occupied in either of two ways: (1) By moving to the selected site and expanding into and organizing the area in the same manner as an on-the-spot establishment, or (2) by halting near the selected site and sending forward reconnaissance forces. The method used must be thoroughly planned and rehearsed. Use of patrol base drills in these methods will assist in swiftly and efficiently establishing patrol bases. Occupation, using an example patrol base drill and subsequent operation, is described in Figures 1-6 and 1-7.

Approach. The patrol leader halts the patrol at a suitable position within 200 meters of the tentative patrol base location. He establishes close-in security. Previously designated individuals (preferably leaders of the patrol's major subdivisions) join the patrol leader (Figure 1-6A).

Reconnaissance. The patrol leader designates the point of entry into the patrol base location as 6 o'clock, assigns areas by the clock system, designates the center of the base as patrol headquarters, and moves to the patrol headquarters. Subordinate leaders reconnoiter assigned areas for suit ability and return to the patrol leader (Figure 1-6B). The patrol leader sends two men to bring the patrol forward.

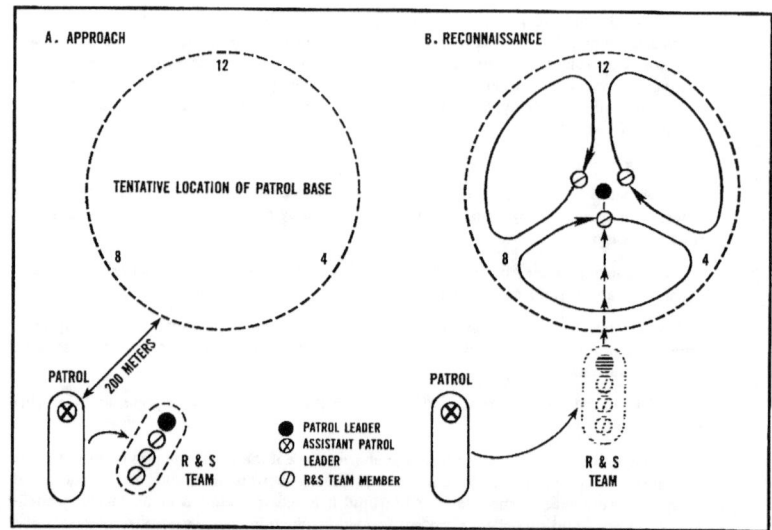

Figure 1-6. Patrol base establishment—reconnaissance.

Occupation. After the tentative patrol base has been reconnoitered, the patrol moves forward to the center of the base as shown in Figure 1-7.

Once the patrol base is established, each unit reconnaissance and security leader reconnoiters his sector and reports indications of enemy or civilians, suitable OP and listening post (LP) positions, rallying points, and withdrawal routes.

The patrol leader designates rallying points, positions for OPs and LPs, and withdrawal routes.

Each unit puts out one 2-man OP (day) and one 3-man LP (night) and establishes communications.

Operations. Make details of the planned *operations* known to all men without assembling all at one time and thus endangering the security of the base.

Limit rehearsals to terrain models, with part of the patrol rehearsing while the remainder provides security. If part of the patrol is absent, adjust the perimeter, if necessary, to ensure security. Do not test-fire weapons.

Ensure orders are as brief as possible. Make maximum practical use of fragmentary orders and references to standing operating procedures (SOPs).

Security is of prime importance during operations. The following principles are vital to the security of every man moving in and around the patrol base.

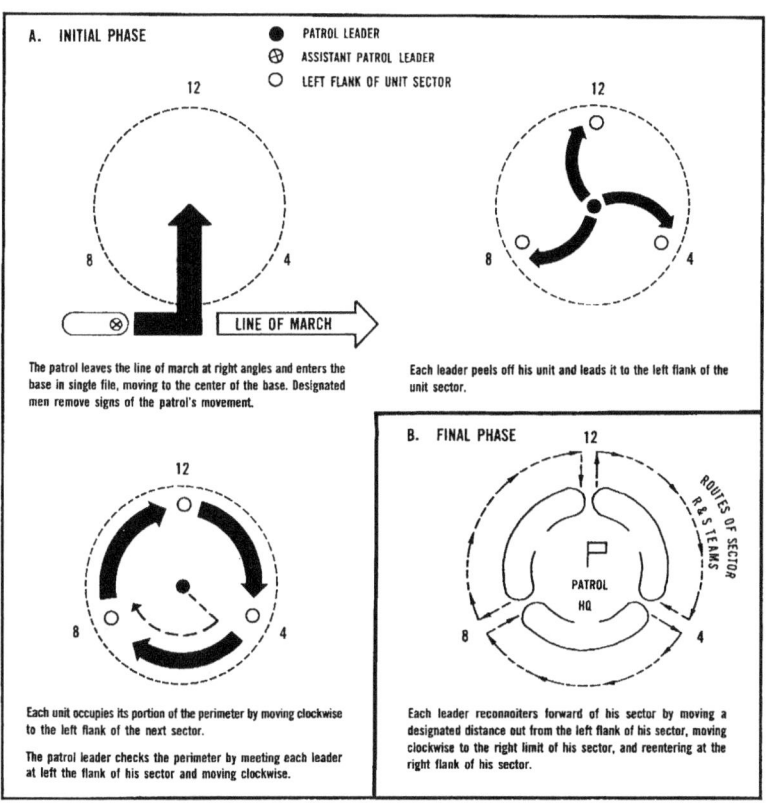

Figure 1-7. Patrol base established—occupation.

Use only one point of entry and exit. Camouflage and guard it at all times.

Build fires only when necessary and, as a general rule, only in daylight. Whether day or night, keep them as small as possible. Where terrain permits, build fires in pits and, if built at night, carefully cover and shield them. Building fires in pits reduces the danger of visual detection and eases extinguishing fires and camouflaging the sites. Use the driest and hardest wood available (to reduce smoke). In most areas, the best time to build fires is when the air is thin and smoke dissipates quickly (usually around noon). Early morning may be appropriate, however, in areas where there is ground fog, weigh the risk of detection because of lingering odor against the risk of detection due to visible smoke.

1-15

Accomplish noisy tasks, such as cutting branches (only at designated times) as early as possible after occupation but never at night or during the quiet periods of early morning and late evening. When possible, perform noisy tasks when other sounds will cover them, such as the sounds of aircraft, artillery, or distant battle noises.

Restrict movement, both inside and outside the patrol base to the minimum necessary.

Detain civilians who discover the location of the patrol base until the base is moved or until they can be evacuated to higher headquarters. Take care to prevent the detained civilians from learning about the base, its operations, and future plans. If necessary, tie them up, blindfold them, and cover their ears.

When sufficient personnel are available, man LPs with at least two, preferably three, individuals so they too can alternate and remain alert.

Observe a one-hour stand-to morning and evening: 30 minutes before and 30 minutes after light in the morning, and 30 minutes before and 30 minutes after dark in the evening. This ensures that every man is acclimated to changing light conditions, and is dressed, equipped, and ready for action.

Make certain each man knows the locations of men and positions to his flanks, front, and rear, and that he knows the times and routes of any expected movement within, into, and out of the patrol base.

A patrol base is usually defended only when evacuation is not possible, but *defensive measures* should be planned.

Do not construct elaborate firing positions.

Stress camouflage and concealment.

Plan for artillery and mortar fires if available. Place early warning devices on avenues of approach. If the base will definitely be defended, place mines, trip flares, and booby traps on avenues of approach and in areas that cannot be covered by fire. The value of these devices must be weighed against the fact that their discovery automatically compromises the patrol base.

Establish an alert plan. This includes plans for evacuation and defense. All members must know these plans and the signals or orders for their implementation. Plans for defending the base include pursuit and destruction of the attacking force.

Establish *communications* with higher headquarters, subordinate units, OPs, and LPs (Figure 1-8). The system must provide for every man to be alerted quickly and quietly.

Radios are an excellent means, but must be carefully controlled.

Wire can be used within the patrol base if its bulk and weight and the time required to lay it and pick it up are not disadvantages.

Tug or pull wires may be used for signaling. They are quiet and reduce radio or telephone traffic.

Messengers may be used within the patrol base.

Perform *maintenance* on weapons and equipment as required.

1-16

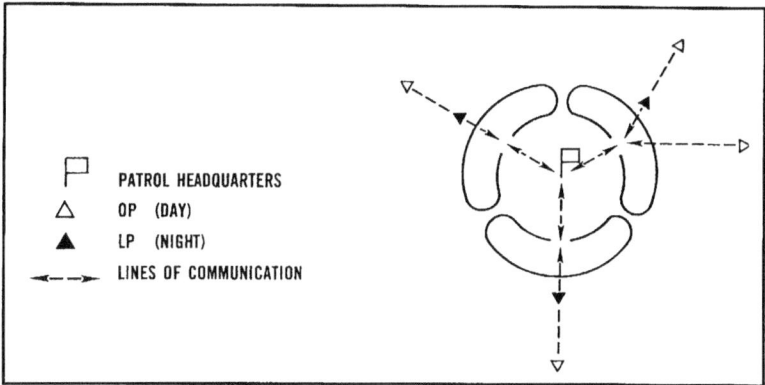

PATROL HEADQUARTERS

OP (DAY)

LP (NIGHT)

LINES OF COMMUNICATION

Figure 1-8. Typical patrol base layout—communications.

Attend to *personal hygiene,* such as washing, shaving, and brushing teeth as needed, and consistent with the situation (including the availability of water). In daylight, use cat holes outside the perimeter. The user must be guarded. At night, locate catholes inside the perimeter. Bury cans, food, and other trash and conceal the burial site.

Eat at staggered times, as planned and controlled. Plan for the *preparation of meals,* if preparation is required.

Guarded *water* parties provide water. Lone individuals do not visit the water source. Make no more than two visits to the source in a 24-hour period. Control the use of water as closely as required.

Permit *rest* and *sleep* only after all work is done. Stagger rest periods so that proper security is maintained. Consistent with work and security requirements, schedule as much sleep and rest as possible for each man.

If the patrol is to be *resupplied* by air, locate the flight path, drop and/or landing zone, and cache so that neither the base nor possible objectives are compromised.

Departing a Patrol Base

Remove or conceal all signs of the patrol's presence. This may prevent the enemy from learning of the patrol's presence in the area, may prevent pursuit, and may prevent the enemy from learning the patrol's methods for operating patrol bases. If possible, avoid night evacuation (in case of attack). When possible, the patrol evacuates as a unit.

TRACKING

A combat worthy tracker must develop and refine certain traits and qualities.

Qualities of a Tracker

A tracker must have patience. He must move slowly, quietly, and steadily while observing and interpreting available indicators. He must avoid using reckless speed that may

1-17

cause him to overlook important signs, lose the trail completely, or blunder into an enemy force. A good tracker must be persistent. He must have the ability and the desire to continue his mission even though indicators are scarce or conditions of weather or terrain are difficult. If he loses the trail, he must have the determination and persistence to find it again. He must be keenly observant to see things that are not obvious at a glance (Figure 1-9). He must be able to use his senses of smell and hearing to supplement his observation. A good tracker must also develop his sixth sense. He may inspect an area simply because it *does not look right*. This ability often enables a tracker to regain a lost trail or discover new or additional indicators. An effective tracker must know the enemy he is fighting. He cannot properly interpret the indicators he has found unless he has some knowledge of the enemy, its habits, equipment, or level of training. A good tracker must also have an understanding of nature, a good memory, intelligence, and he must be physically able to accomplish his mission.

Concepts of Tracking

As a tracker moves along the ground following a trail, he begins to build a picture in his mind of the enemy he is following. To do this, he asks himself the following questions: How many persons am I following? What is their state of training? How are they equipped? Are they healthy? What is their state of morale? Do they know they are being followed? To answer these questions, the tracker uses available indicators, that is, anything that tells a certain action took place at a particular place and time. For instance, a footprint in soft sand is an excellent indicator. The tracker can determine a particular time a person walked on this spot.

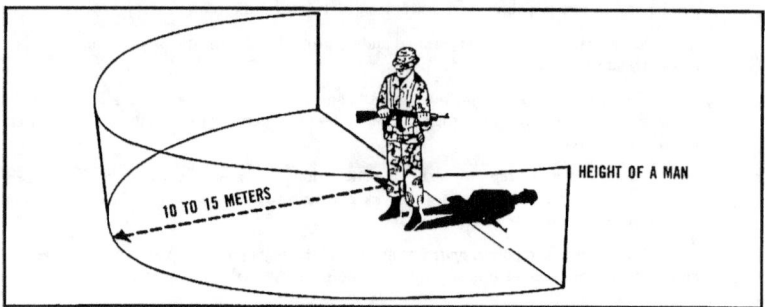

Figure 1-9. Area surveyed by tracker for indicators.

By comparing indicators, the tracker gets answers to his questions. If he finds a footprint and a scuff on a tree about waist high, for example, it may indicate that an armed man passed this particular spot.

Six concepts apply to tracking. Any indicator that the tracker discovers can be defined by one or more of the following concepts. They are—

- Displacement.
- Stains.
- Weather.

1-18

32

- Litter.

- Camouflage.

- Immediate use intelligence.

For the sixth concept, the tracker combines all indicators and interprets what he has seen to form a composite picture. He can then use this composite picture as on–the–spot intelligence. For example, indicators may indicate contact is imminent and extreme stealth may be required.

Displacement

Displacement takes place when anything is moved from its original position. A well–defined footprint in soft, moist ground is a good example of displacement. The shoe or foot of the individual who left the print displaced the soil by compression, thus leaving the indentation in the ground. By studying this indicator, the tracker can determine several important facts. The print left by worn footwear or by a barefooted person may indicate lack of proper equipment.

Analyzing footprints. Footprints may indicate direction and rate of movement, number of persons in the moving party, whether or not heavy loads are being carried, sex of members of the party, and whether the members of the party realize that they are being followed (Figure 1–10).

Determining key prints. Since the last man in a file normally leaves the clearest footprints, his should be the *key* set of prints. The tracker should cut a stick to match the length of the key prints and notch it to indicate the width at the widest part of the sole. He should study the angle of the key prints to the direction of march. He should also look for an identifying mark or feature on the prints, such as a worn or frayed part of footwear, to help him identify the key prints. In case the trail becomes vague or erased, or merges with another, the tracker can use his stick measuring devices and with close study can identify the key prints. This will aid him to stay on the trail. By using the box method, the tracker can count up to 18 persons. There are two ways the tracker can employ the box method.

The first and most accurate way is to use the *stride* as a unit of measure when key prints can be determined. The tracker uses the set of key prints and the edges of the road or trail to box in an area to analyze (Figure 1–11A).

The second way a tracker may use the box method is the *36–inch box*. The 36–inch box is used when there are no key prints distinguishable. However, this system is not as accurate as the stride measurement (Figure 1–11B).

Recognizing other signs of displacement. Footprints are only one example of displacement. Anything that has been moved from its original position by a moving person is an example of displacement.

Foliage, moss, vines, sticks, or rocks that are scuffed or snagged from their original place form valuable indicators (Figure 1–12A). Vines may be dragged, dew droplets may be displaced from leaves, or stones and sticks may be turned over to indicate a different color underneath. Grass or other vegetation may be bent or broken in the direction of movement.

Bits of clothing, threads, or dirt from boots can be displaced from a person's uniform and left on thorns, on snags, or on the ground. The tracker should inspect all areas for bits of clothing or other matter ripped from the uniform of the person being tracked.

1–19

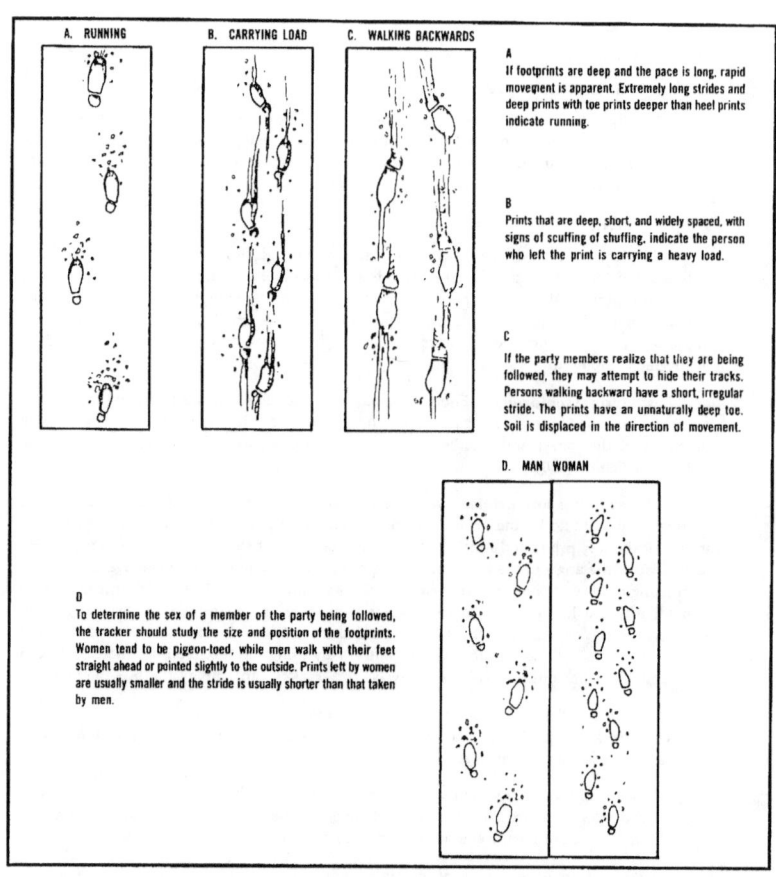

Figure 1-10. Different types of footprints.

Wild animals and birds flushed from their natural habitat by the person being tracked is another example of displacement. Cries of birds excited by unnatural movement is an indicator. Tops of tall grass or brush moving on a windless day indicates that someone is moving the vegetation.

Displacement can result from clearing a trail by either breaking or cutting one's way through heavy vegetation with a machete. Such trails are obvious to the most inexperienced tracker. Some individuals may unconsciously break additional branches as they move behind the person who is cutting. Displacement indicators can be made by persons carrying heavy

1-20

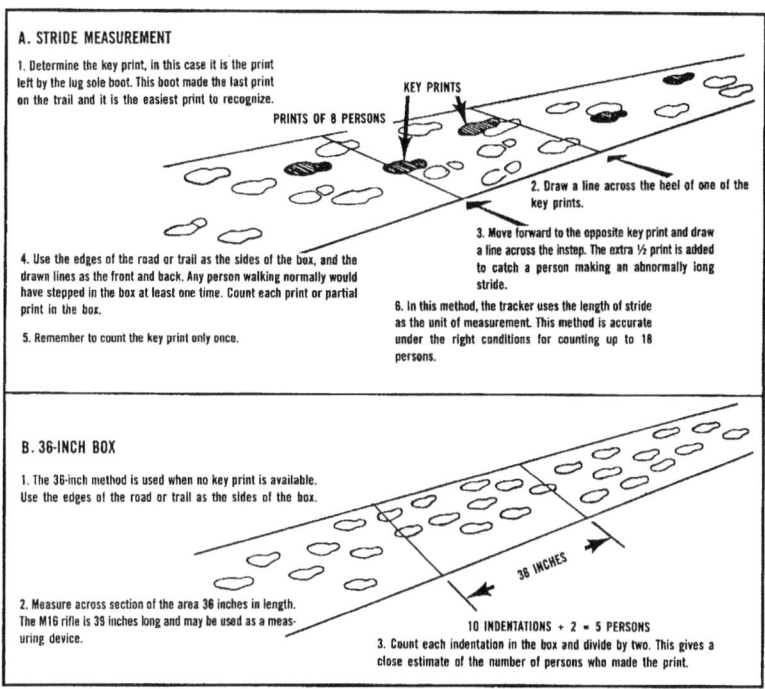

A. STRIDE MEASUREMENT

1. Determine the key print, in this case it is the print left by the lug sole boot. This boot made the last print on the trail and it is the easiest print to recognize.

KEY PRINTS

PRINTS OF 8 PERSONS

2. Draw a line across the heel of one of the key prints.

3. Move forward to the opposite key print and draw a line across the instep. The extra ½ print is added to catch a person making an abnormally long stride.

4. Use the edges of the road or trail as the sides of the box, and the drawn lines as the front and back. Any person walking normally would have stepped in the box at least one time. Count each print or partial print in the box.

5. Remember to count the key print only once.

6. In this method, the tracker uses the length of stride as the unit of measurement. This method is accurate under the right conditions for counting up to 18 persons.

B. 36-INCH BOX

1. The 36-inch method is used when no key print is available. Use the edges of the road or trail as the sides of the box.

2. Measure across section of the area 36 inches in length. The M16 rifle is 39 inches long and may be used as a measuring device.

36 INCHES

10 INDENTATIONS ÷ 2 = 5 PERSONS

3. Count each indentation in the box and divide by two. This gives a close estimate of the number of persons who made the print.

Figure 1-11. Box methods of determining number of footprints.

A. TURNED OVER ROCKS AND STICKS

B. CRUSHED AND DISTURBED VEGETATION

C. SLIP MARK AND WATERFILLED FOOTPRINTS ON STREAM BANKS.

Figure 1-12. Examples of displacement.

1-21

loads stopping to rest. Prints made by the edges of boxes can help to identify the load carried. When loads are set down at a rest halt or campsite, grass and twigs are usually crushed (Figure 1-12B). A man reclining will also flatten the vegetation.

In almost any area, there are insects and spiders. The observation of any changes in the normal life of these insects may indicate that someone has recently passed. Bees that are stirred up, ants that have had their holes covered by someone moving over them, or spiders that have had their webs torn down are valuable clues. Spiders often spin webs across open areas, trails, or roads to trap flying insects. If the person being followed is careless and does not move under these webs, he is leaving an indicator to an observant tracker.

If the person being followed attempts to use a stream to cover his trail, the tracker may still be able to follow successfully. The person may displace algae and other water plants by losing his footing or by walking carelessly. Rocks may be displaced from their original position or turned over to indicate a lighter or darker color on the opposite side. The person entering or exiting a stream may create slide marks, footprints, or scuff bark off of roots or sticks (Figure 1-12C). Normally a person or animal seeks the path of least resistance; therefore, when searching the stream for indication of departures, open areas along the banks may often yield results.

Stains

A stain occurs when any substance from one organism or article is smeared or deposited on something else. The best example of staining is blood from a profusely bleeding wound. Bloodstains are often in the form of spatters or drops. Blood indicators are not always on the ground. Blood may also be smeared on the leaves or twigs of trees and bushes.

By studying the bloodstains left by a wounded person, the tracker can determine the location of the wound on the person. For example—

- If the blood seems to be dripping steadily, it probably came from a wound on the trunk.

- If the blood appears to be slung to the front, rear, or to the sides, the wound is probably in an extremity.

- Arterial wounds appear to pour the blood at regular intervals as if it were poured from a pitcher. If the wound is venous, the blood pours steadily.

- A wound in a lung deposits bloodstains that are pink, bubbly, and frothy.

- A bloodstain from a head wound appears heavy, wet, and slimy, like gelatin.

- Abdominal wounds often mix blood with digestive juices so the deposit has an odor and is light in color.

The tracker can also determine the seriousness of the wound and how far the wounded person can move unassisted. This process may lead the tracker to enemy bodies or may indicate where they have been carried.

Staining can also occur when muddy footgear is dragged over grass, stones, and shrubs. Thus, staining and displacement combine to indicate movement and direction. Crushed leaves may stain rocky ground that is too hard to leave footprints. Roots, stones, and vines may be stained where leaves or berries are crushed on them by moving feet.

In some instances it may be hard to determine the difference between staining and displacement. Both terms can be applied to some indicators. For example, water that has been muddied may indicate recent movement. Mud that has been displaced also stains the water. Stones in streams may be stained by mud from footwear. Algae can be displaced from stones in streams and can stain other stones or the bank.

Water that collects in footprints in swampy ground is muddy if the tracks are recent. With time, however, the mud settles and the water clears. The tracker can use this information to indicate time. Normally, the mud clears in approximately one hour. Clearing time, of course, varies with the terrain.

Weather

Weather can either aid or hinder the tracker. It affects indicators in certain ways so that the tracker can determine their relative ages. Wind, snow, rain, or sunlight may, however, erase indicators entirely, thus hindering the tracker.

By studying the effects of weather on indicators, the tracker can determine the age of the sign. For example, when bloodstains are fresh, they are bright red. Air and sunlight change blood first to a deep ruby-red color, then to a dark brown crust when the moisture evaporates. Scuff marks on trees or bushes darken with time, sap oozes, then hardens when it makes contact with the air.

Footprints are greatly affected by weather (Figure 1-13). By carefully studying this weathering process, the tracker can determine the approximate age of the print. If particles are just beginning to fall into the print, the tracker should become a stalker. If the edges of the print are dried and crusty, the prints are probably at least an hour old. This varies with terrain and should be taken as a general guide.

A light rain may round out the edges of the print. The tracker must remember when the last rain occurred in order to place the print into a proper time frame. A heavy rain may erase all signs.

Wind also affects tracks. Besides drying out the tracks, litter, sticks, or leaves may be blown into the prints. By remembering wind activity, the tracker may guess the age of the tracks. For example, the tracker may think, *It is calm at the present, but the wind blew hard about an hour ago. These tracks have litter blown into them, so they must be over an hour old.* The tracker must be sure, however, that the litter was blown into the prints, and not crushed into them when the prints were made.

Trails exiting streams may appear to have been weathered by rain because of water running from clothing or equipment into the tracks. This is particularly true if the party exits

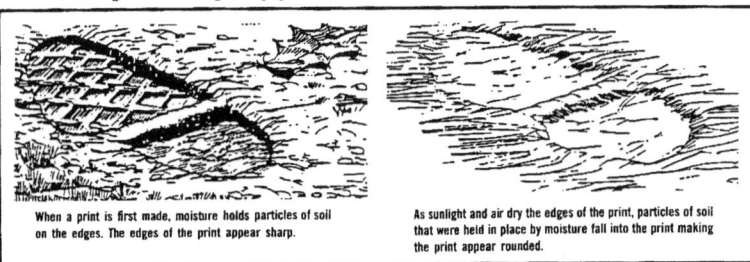

When a print is first made, moisture holds particles of soil on the edges. The edges of the print appear sharp.

As sunlight and air dry the edges of the print, particles of soil that were held in place by moisture fall into the print making the print appear rounded.

Figure 1-13. Weather effects on footprints.

1-23

the stream in a file. This will permit each person to deposit water into the tracks. The existence of a wet, apparently weathered trail slowly fading into a dry trail indicates the trail is fresh.

Wind affects sounds and odors. If the wind is blowing from the direction of the trail that the tracker is following, sounds and odors may be carried to him. If the wind is blowing in the same direction as the trail being followed, the tracker must be extremely cautious since the wind will carry his sounds toward the enemy. The tracker can determine wind direction by dropping a handful of dry dust or grass from shoulder height. By pointing in the same direction the wind is blowing, the tracker can localize sounds by cupping his hands behind his ears and slowly turning. When the sounds are loudest, the tracker is facing the origin of the sound.

In calm weather, when no wind is blowing, the air currents that may carry sounds to the tracker may be too light to feel. The tracker must remember that the air cools in the evening and moves downhill to the valleys. If the tracker is moving uphill late in the day or at night, air currents will probably be moving toward him, provided no other wind is blowing. As the sun warms the air in the valleys in the morning, it moves uphill. These factors should be considered when plotting routes for patrols or other operations. If the tracker can keep the wind in his face, sounds and odors will be carried to him from his objective or the party being tracked.

The sun should also be considered by the tracker. It is difficult to shoot directly into the sun. If the tracker has the sun at his back and the wind in his face, he has a slight advantage.

The tracker should know and understand how weather affects soil, vegetation, and other indicators in his area. He cannot properly determine the age of indicators until he understands the effects that weather has on trail signs.

Litter

A poorly trained or poorly disciplined unit moving over a piece of terrain is likely to leave a clear trail of litter. Gum or candy wrappers, ration cans, cigarette butts, remains of fires, or even piles of human feces are unmistakable signs of recent movement. The tracker must consider weather when estimating the age of litter. Rain flattens or washes litter away and turns paper into pulp. Ration cans, exposed to weather, will rust first at the exposed edge where it is opened. Rust then moves in toward the center. Again the tracker must use his memory to properly determine the age of litter. The last rain or strong wind can be the basis for a time frame.

Camouflage

Camouflage applies to tracking when the party being followed employs techniques to baffle the tracker or slow him down. Walking backward to leave confusing prints, brushing out trails, and moving over rocky ground or through streams are examples of techniques that may be employed to confuse the tracker. The party being followed may employ *most used* and *least used* routes to cover their movement.

Movement on lightly traveled sandy or soft trails is easily trailed. However, a guerrilla may attempt to confuse the tracker by moving on hard-surfaced, frequently traveled roads. He may even attempt to merge with the civilians. These routes should be examined carefully because if a well-defined approach leads to the enemy, it will probably be mined, ambushed, or covered by snipers.

1-24

Least used routes are used to confuse the tracker by avoiding all man—made trails or roads. These routes are normally magnetic azimuths between two points. The tracker can, however, by using the proper concepts, follow the party if he is experienced and persistent.

The party being followed may use several methods to minimize trail signs. Footwear wrapped with rags or soft—soled sneakers may make footprints more rounded on the edges and less distinct. The party may exit a stream in a column or a line; this reduces the chance of leaving a well—defined exit.

By studying signs, a careful, observant tracker can determine if an attempt is being made to confuse him. If the party attempts to throw off the tracker by walking backward, the footprints will be deepened at the toe and soil will be scuffed or dragged in the direction of movement. By following carefully, the tracker can normally find a turnaround point.

A trail can be brushed out, but rarely without leaving signs. The experienced tracker can easily recognize this technique.

If the trail leads across rocky or hard ground, the tracker should circle the area to pick up the exit trail. On rocky ground, moss or lichens growing on the stones can be displaced by even the most careful evader. If these methods fail, the tracker should return to the last visible indicators and then head in the direction of movement in ever-widening circles until he again falls upon visible signs.

Remember that anyone who attempts to hide his trail moves at a reduced speed. Therefore, the experienced tracker, who is not fooled by these attempts, gains time on the person being followed.

Immediate Use Intelligence

As the tracker moves along, he constantly asks himself questions. As he finds indicators that answer those questions, he begins to form a picture of the enemy in his mind.

Interpreting. The tracker must avoid reporting his interpretations as facts. He should report that he has seen indications of certain things instead of stating to the commander that these things actually exist. The commander may have additional information to help him estimate the enemy he is facing.

There are many ways a tracker can make interpretations, as discussed above, relating to size of the party, sex, load, equipment, and many more things. Time frames can be determined by the effects of weather on the indicators.

Reporting. Immediate use intelligence is information concerning the enemy that can be put to use immediately to gain surprise, to keep the enemy off balance, or to keep it from escaping the area entirely. The commander may have many sources of intelligence: reports, documents, prisoners of war. These sources can be put together to form indications of where the enemy was, what it may be planning, and where it may be going.

Tracking, however, gives the commander definite information on which he can act immediately. For example, a unit may report that there are no men of military age in a village. This information is of value if it is combined with other information to make a composite enemy picture in the area. A tracker, however, who interprets trail signs and reports to his commander that he is 30 minutes behind a known enemy unit, that he is moving north, and that he is located at a particular location, gives the commander information on which he can act at once.

Tracking, therefore, is one of the best sources of immediate use intelligence. Indicators may be so fresh that the tracker becomes a stalker, or they may provide information that will allow the commander to plan a successful operation.

INTERDICTION OPERATIONS

Interdiction operations conducted against an enemy can hinder or interrupt lines of communication; deny use of certain key areas; and destroy industrial facilities, military installations, equipment and resources. Mines (to include booby traps) and snipers can be used to interdict enemy targets. These methods of interdiction may also be used to support raids and ambushes. For a detailed discussion on targeting, see SFOD-I Student Handbook on Special Operations Targeting, editions 1 through 4.

Mining

Mining affords the SF and resistance force commanders a means of interdicting enemy routes of communication and key areas with little manpower expense. Mines may be used in support of specific tactical operations or for general harassment of the enemy by emplacement along routes of enemy movement. They may be emplaced around installations to cause casualties, limit movement, and induce low morale among enemy troops. For detailed information on the use and installation of mines, booby traps, and other devices, see (C) FM 5-31.

Sniping

Sniping as an interdiction technique has a demoralizing effect on the enemy. A few well-trained and properly deployed snipers can cause numerous casualties. They can hinder or temporarily deny enemy use of certain routes or areas. Sniping operations may require the enemy to employ a disproportionate number of troops to rid the area of snipers. Detachment commanders selecting, training, and deploying snipers must be completely familiar with their use and be able to train them properly. They must also plan for logistical support in acquiring special sniper equipment to make snipers effective in all types of operations.

The sniper. A sniper is an expert rifleman, physically and mentally hardened to endure long periods of loneliness and hardship. He must be able to—

- Estimate ranges.
- Search areas systematically.
- Locate and identify sounds.
- Use cover, concealment, and camouflage.
- Use maps, sketches, aerial photographs, and the compass.
- Recognize enemy personnel and equipment quickly.
- Move without detection.
- Endure long periods of waiting (patience).

Missions. Snipers assigned areas of responsibility should have mission-type orders outlining priority targets that may include killing key enemy personnel such as patrol leaders, gunners of crew-served and automatic weapons, communication personnel, observers, and enemy snipers. In the absence of these priority targets, they may fire on any enemy personnel. Snipers may cover an area that has been mined to prevent removal or exploitation of the minefield. They may be used as part of a raid or ambush to stop enemy personnel escaping the area under attack. In addition to their sniping mission, they may collect information for intelligence sections of the area command or guerrilla units. In their

1-26

constant search for targets, they become thoroughly familiar with the terrain, enemy actions, and movements, routes of communications, and other activities.

Selection and training. Detachment commanders and resistance force commanders select snipers from their outstanding guerrilla force personnel, specifically the rifleman in operational units. Additional training should be given in maintenance and operation of electronic night firing devices, viewing devices such as telescopic sights, and other types of firing devices as the commanders deem necessary or as time permits.

Sniper employment planning. Plans must be made to properly locate a single sniper or sniper teams. Area commanders should incorporate the use of snipers into the tactical plan, and coordinate their use with subsector commanders. When snipers are employed in specific areas, all operations should be curtailed in that area or conducted on a limited basis. Special provisions must be made for snipers to rest and recuperate after strenuous tours of duty. Sniper employment may require a special unit, tightly controlled by the area command.

Sniper teams. Snipers are best employed in pairs, particularly when operating from a stationary post. Remaining in one position for long periods of time, and using binoculars constantly, places a heavy strain on one man. By working in pairs, snipers can alternate duties, thus keeping their post in continuous operation. One observes and estimates ranges, while the other fires. The first shot should be a hit.

The single sniper. A single sniper is normally used when two might be detected. He can often cover a large area by moving from one firing position to another as often as required in search of worthwhile targets and good fields of fire. Close coordination between sector and subsector commanders is required in these cases.

Equipment. The sniper carries only the equipment and supplies needed to complete the mission within an estimated time. In some instances, he may have to rely on MSSs or caches to replenish supplies and equipment for either his operational role or his survival. The decision to release the location of these sites to a sniper rests with the area commander. The sniper may need, as a minimum—

- His weapon, telescope sight and, if available, infrared weapon sight or ametascope.
- Binoculars.
- Watch.
- Compass.
- Map.
- Camouflaged clothing.
- Ammunition.
- Individual rations.

NOTE: Other equipment to support assigned missions should be obtained as required. (For more information on sniper training and employment see TC 23-14.)

Section II. Defensive Operations

This section describes those defensive tactics and techniques the SF soldier must know to successfully defend himself and his team members in support of mission requirements. Subjects discussed include guerrilla operations, indicators of counterguerrilla operations, defensive tactics, counterambush, countertracking, as well as breakout operations.

1-27

GUERRILLA OPERATIONS

Guerrilla units are normally inferior to organized enemy forces in strength, firepower, mobility, and communications. Therefore, guerrilla operations are primarily offensive; they do not undertake defensive operations unless forced or ordered to. Usually, when the enemy attacks, guerrillas defend themselves by movement, dispersion, withdrawal, or diversions. Defensive operations are accompanied whenever possible by offensive actions against the enemy's flanks and rear. On this basis, SF always plan offensive operations and security within the UWOA.

INDICATORS OF COUNTERGUERRILLA OPERATIONS

Security of the UWOA requires guerrilla intelligence measures to identify indications of impending counterguerrilla action, population control measures, and guerrilla reaction to enemy counterguerrilla actions. Some activities and conditions that may indicate impending enemy counterguerrilla actions are—

- Suitable weather.
- New enemy commander.
- Changes in battle situation elsewhere.
- Arrival of new enemy units with special training.
- Extension of enemy outposts, increased patrolling, and aerial reconnaissance.
- Increased enemy intelligence effort.
- Civilian pacification or control measures.
- Increased PSYOP against guerrillas.

Some measures that may be used to control the population of an area are—

- Mass registration.
- Curfews.
- Intensive propaganda.
- Compartmentalization, with cleared buffer zones.
- Informer nets.
- Party membership drives.
- Land and housing reform.
- Relocation of individuals, groups, and towns.
- Rationing of food and goods.

DEFENSIVE TACTICS

The existence or indication of counterguerrilla operations requires the SF and guerrilla force commanders to plan and use defensive tactics. Discussed below are some of the defensive tactics applicable against counterguerrilla operations.

Diversion Activities

A sudden increase in guerrilla activities or a shift of such activities to other areas assists in diverting enemy attention. For example, intensified operations against enemy lines of

1-28

communications and installations require the enemy to divert troops from counterguerrilla operations to security roles. Full use of underground and auxiliary capabilities assists in creating diversions.

Defense of Fixed Positions

The rules for a guerrilla defense of fixed positions are the same as those for conventional forces, except there are few supporting fires and counterattacks are generally not practicable. In conjunction with their position defense, elements of the guerrilla force conduct raids, ambushes, and attacks against the enemy's lines of communication, flanks, reserve units, supporting arms, and installations. Routes of approach are mined and camouflaged snipers engage appropriate enemy targets. Diversionary actions by all elements of the resistance movement are increased in adjacent areas.

Delay and Harassment Activities

The objective of delay and harassment tactics is to make the attack so costly that the enemy eventually ends its operations. Defensive characteristics of the terrain are used to the maximum, mines and snipers are employed to harass the enemy, and ambushes are positioned to inflict maximum casualties and delay.

As the enemy overruns various strong points, the guerrilla force withdraws to successive defensive positions to again delay and harass. When the situation permits, the guerrilla force attacks the enemy's flanks, rear, and lines of communication. If the enemy continues its offensive, the guerrilla forces should withdraw and leave the area. Under no circumstances should the guerrilla force become so engaged that it loses its freedom of action and permits enemy forces to encircle and destroy it.

Withdrawal

In preparing to meet enemy offensive action, the SF and guerrilla force commanders may decide to withdraw to another area not likely to be included in the enemy offensive. Key installations within a guerrilla base are moved to alternate bases, and essential records and supplies may be transferred to new locations. Less essential items will be destroyed or cached in dispersed locations. If the commander receives positive intelligence about the enemy's plans for a major counterguerrilla operation, he may decide to withdraw and leave his main base without delay.

When faced with an enemy offensive of overwhelming strength, the commander may disperse his force in either small units or as individuals to avoid destruction. This course of action, however, renders the guerrilla force ineffective for an undetermined period of time and therefore should not be taken unless absolutely necessary.

COUNTERAMBUSH

The very nature of ambush, "...a surprise attack from a concealed position," places the ambushed unit at a disadvantage. Obviously, the best defense is to avoid being ambushed, but this is not always possible. A unit must, therefore, reduce its vulnerability to ambush and reduce the damage it will sustain if ambushed. These measures must be supplemented by measures to destroy or escape from an ambush.

Reduction of Vulnerability to Ambush

No single defensive measure or combination of measures can prevent or effectively counter all ambushes in all situations. The effectiveness of counterambush measures is directly related to the state of training of the unit and the leadership ability of its leader.

1-29

In avoiding ambush, dismounted units have an advantage over mounted units. They are less bound to the more obvious routes of movement, such as roads and trails (as in armored units).

Dismounted units are at a disadvantage, however, when—

- Terrain, such as heavy jungle, restricts or prohibits cross–country movement.
- The need for speed requires movement on roads, trails, or waterways.

Preparation for Movement

In preparing for movement, the leader must consider the terrain and all information he has on the enemy. In doing so, he studies maps of the area and if possible, makes an aerial reconnaissance.

Map. In studying maps of the terrain over which the leader will move his unit, the leader first checks the map's marginal data to determine reliability at the time the map was made. If reliability is not good, or if the map is old, he evaluates its reliability in light of all other information he can obtain. For example, a 20–year old map may not show several nearby roads and trails; more recent building development in the area will not be shown. The leader considers the terrain in relation to all available information of known or suspected enemy positions and previous ambush sites. His map study includes evaluation of the terrain from the enemy's viewpoint. How could the enemy use this terrain? Where could the enemy position troops, installations, and ambushes?

Aerial reconnaissance. If possible, the leader makes an aerial reconnaissance. The information gained from the aerial reconnaissance enables him to compare the map and the terrain. He also obtains current and more complete information on roads, trails, man–made objects, type and density of vegetation, and seasonal condition of streams.

An aerial reconnaissance also reveals—

- Movement or lack of movement in an area (friendly, enemy, civilian).
- Indications of enemy activity. Smoke may indicate locations of campsites, patrols, or patrol bases. Freshly dug soil may indicate positions or ambush sites. Shadows may aid in identifying objects. Unusual shapes, sizes, shadows, shades, or colors may indicate faulty camouflage.

Despite its many advantages, aerial reconnaissance has limitations. For example—

- Strength of bridges cannot be determined.
- Terrain surface may be misinterpreted.
- Mines and booby traps cannot be seen.
- Presence of aircraft may warn enemy.

Route selection. The factors the leader considers are the same whether he is selecting a route or studying a route he has been directed to follow.

Cover and concealment. Cover and concealment are desirable, but a route with these features may obstruct movement. Terrain that provides a moving unit cover and concealment also provides the enemy increased opportunities for ambush. Identification of areas where ambushes may be concealed allows the leader to develop plans for clearing these areas.

How the terrain affects *observation* and *fields of fire* available to the unit and to the enemy will influence the selection of and movement over a route, formations, rates of movement, and methods of control.

Key terrain. Key terrain is an earth feature that has a controlling effect on the surrounding terrain. It must be identified and actions planned accordingly. If, for example, a hill provides observation and fields of fire on any part of a route, the leader must plan for taking the hill from the enemy.

Obstacles. Obstacles may impede movement or limit maneuver along a route. They may also limit enemy action.

Current intelligence. All available information is considered. This includes—

- Known, suspected, and previous ambush sites.
- Weather and light data.
- Reports of units or patrols that have recently operated in the area.
- Size, location, activity, and capabilities of guerrilla forces in the area.
- Attitude of the civilian population and the extent to which they can be expected to cooperate or interfere.

Counterintelligence. In counterguerrilla operations, in particular, a key feature of preparing for movement is denying the enemy information. A unit is especially vulnerable to ambush if the enemy knows the unit is to move, what time it is to move, where it is to go, the route it is to follow, and the weapons and equipment it is to carry. The efforts made to deny or delay enemy acquisition of this information comprise the counterintelligence plan. As a minimum, the plan restricts dissemination of information.

The leader gives out mission information only on a need-to-know basis. This is especially important when the native personnel operating with the unit might possibly be planted informers. Once critical information is given, personnel are isolated so that nothing can be passed out.

If it is likely that the enemy or enemy informers will observe the departure of a unit, deception plans may be used. Two examples are shown in Figures 1-14 and 1-15.

Communications. The leader plans how he will communicate with elements of his unit; with artillery, air, or other supporting units; and with higher headquarters. On an extended move, a radio relay or a field expedient antenna may be necessary. An aircraft might be used to communicate with air or artillery support.

Fire support. The leader plans artillery and mortar fires so they will deceive, harass, or destroy the enemy. They may be planned as scheduled or on-call fires.

Fires are planned—

- On key terrain features along the route. These can serve as navigational aids or to deceive, harass, or destroy enemy.
- On known enemy positions.
- On known or suspected ambush sites.
- On the flanks of identified danger areas.
- Wherever a diversion appears desirable. For example, if the unit must pass near an identified enemy position, artillery or mortar fires on the position may distract the enemy and permit the unit to pass undetected.

1-31

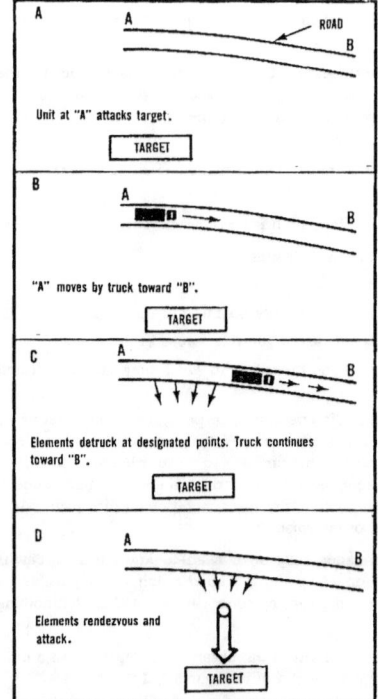

Figure 1-14. Example Deception Plan I.

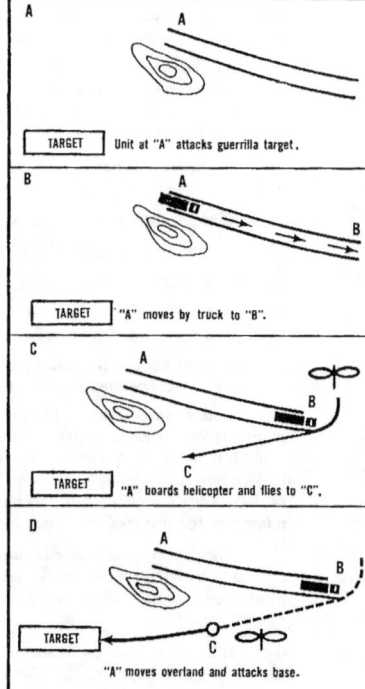

Figure 1-15. Example Deception Plan II.

- At intervals along the route, every 500 or 1,000 meters for example. With fires so planned, the unit is never far from a plotted concentration from which a shift can be quickly made.

Coordination with the supporting unit includes—

- Route to be followed.
- Scheduled and on-call fires.
- Call signs and frequencies.
- Checkpoints, phase lines, and other control measures.
- Times of departure and return.

1-32

46

Organization for Movement

Intelligence. The unit must provide its own intelligence support. Members must be alert to report information, and leaders must be able to evaluate the significance of this information in relation to the situation.

Obvious items from which intelligence may be gained are—

- Signs of passage of groups, such as crushed grass, broken branches, footprints, cigarette butts, or trash. These may reveal identity, size, direction of travel, and time of passage.
- Workers in fields. This may indicate absence of the enemy.
- Apparently normal activities in villages. These may indicate absence of the enemy.

Less obvious items from which *negative information* may be gained are—

- The absence of workers in fields. This may indicate that the enemy is near.
- The absence of children in a village. This may indicate that they have been hidden to protect them from action that may be about to take place.
- The absence of young men in a village. This indicates that the village is controlled by the enemy.

A knowledge of enemy signaling devices is very helpful. Those listed below are some that were used by Communist guerrillas in Vietnam:

- A farm cart moving at night shows one lantern to indicate that no government troops are on the road or trail behind. Two lanterns mean that government troops are close behind.
- A worker in the fields stops to put on or take off his shirt. Either act can signal the approach of government troops. This is relayed by other informers.
- A villager fishing in a rice paddy holds his pole out straight to signal all clear; up at an angle to signal that troops are approaching.

Security. Security is obtained through organization for movement, manner of movement, and by every man keeping alert at all times.

A two-man patrol can maintain security by organizing into a security team with sectors of responsibility as shown in Figure 1-16.

A larger unit can use any standard formation (file, column, vee) and establish a reaction force. This reaction force can be separated from the main body so that it does not come under the first of an ambush of the main body, and can maneuver to destroy the ambush.

Any unit of squad size or larger, regardless of the formation used, must have security forces to the front, flanks, and rear. A rifle company organized in this manner is shown in Figure 1-17.

A dismounted unit moves by the same methods as a motorized patrol—continuous movement, successive bounds, or alternate bounds.

1-33

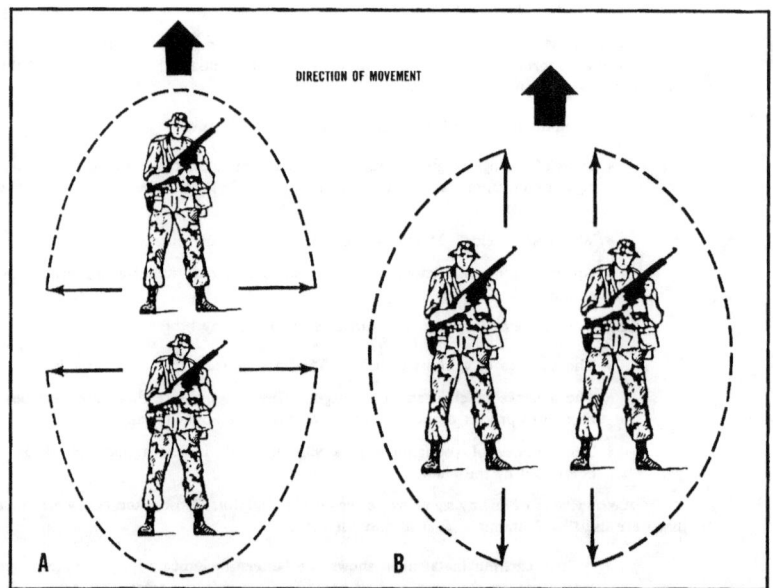

Figure 1-16. Security in a two-man patrol.

Counterambush Immediate Action Drill.

If ambushed, the unit reacts with the appropriate counterambush immediate action drill. (Refer to Appendix B.)

COUNTERTRACKING

To be a more effective tracker and a more effective soldier on patrol, you should know something about countertracking techniques.

Techniques Used to Avoid an Enemy Tracker

If the person tracking you is not an experienced tracker, some of the following techniques may throw him off.

Moving from a thick area to an open area. While moving in any given direction from a thick area to a more open area, walk past a big tree (3/4-foot or larger diameter) toward the open vegetation for three to five paces and then walk backward to the forward side of the tree and make a 90-degree change in direction, passing the tree on its forward side. Step carefully and leave as little sign as possible. If this is not the direction that you wish to go, change direction again 50 to 100 meters farther at another suitably located big tree and repeat the previous steps. The purpose is to draw the following party into the open

1-34

48

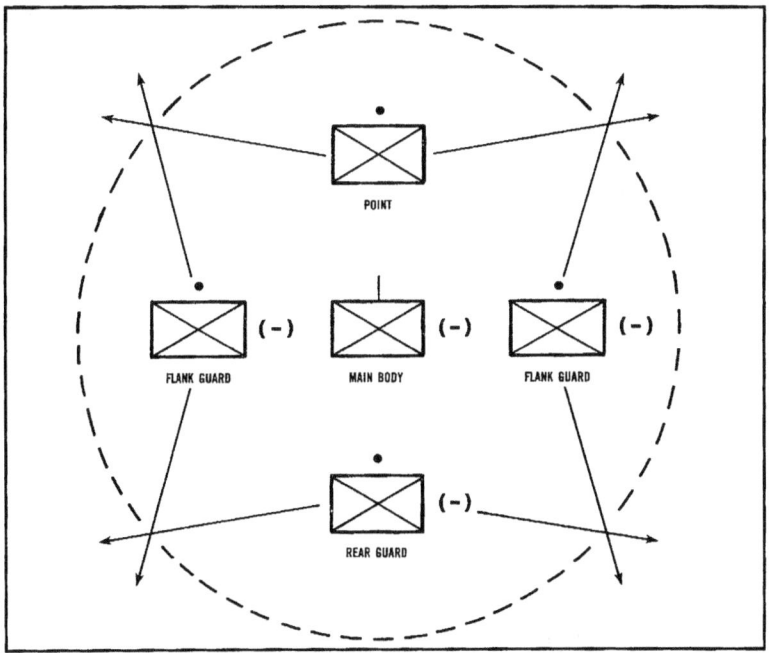

Figure 1-17. Security organization of a rifle company.

area where it is harder for it to track. This maneuver may lead the party to search in the wrong area before it realizes that it has lost the track.

Moving through a known area. Change direction near a marked track when moving through a known area and upon an established jungle track going at right angles to your line of movement. Before reaching the track (100 meters) change direction and approach the track at a 45-degree angle (Figure 1-18). After arrival at the track, continue forward along the track 20 to 30 meters. Leave considerable ground and top signs of your presence. Then, walk backward to the point where you joined the track, go straight across the track and leave no sign of your reentering the jungle. Move off for 100 meters at a 45-degree angle, but this time on the other side of the track and in the reverse of your approach march. Detail the last member of your patrol to cover up all signs of your movement. When changing direction back to your original line of march, use the big tree technique. The purpose of this tactic is to draw the following party along the easier going jungle track. You have, by changing direction before reaching the track, indicated that this is your new line of march. If you are successful, the following party will cast even farther away in the wrong direction before it realizes that it has lost your track.

1-35

49

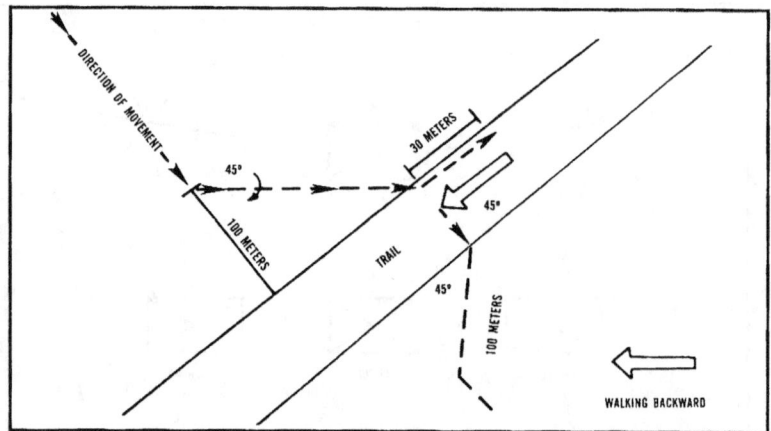

Figure 1-18. Approach trail.

Leaving footprints. Walk backward over soft ground to leave reasonably clear footprints. Try not to leave every footprint clear and do not leave an impression of more than 1/4 inch deep. Continue this deception until you are on hard ground. Select the ground carefully to ensure that you have at least 20 to 30 meters of this deception. This technique is normally used when leaving a stream. The purpose of leaving backward footprints is to get the following party to look in a direction opposite to your line of travel. Always use this technique when coming out of a river or stream. To add even further confusion to the following party, this tactic can be used several times to lay false trails before actually leaving the stream.

Approaching a stream. When moving through a familiar area, change direction 100 meters before approaching a known stream and approach it at a 45-degree angle. When entering the stream, turn in a false direction and proceed down the stream for at least 20 to 30 meters and then backtrack and move off into the intended direction (Figure 1-19). Changing direction before entering the stream can confuse any following party. When the following party enters the stream it should follow the false trail until the track is lost. The following party will be in a false-start position to try and relocate the track. It will begin to probe farther, and will get farther away. It will have to start examining both banks. Following a false trail is time-consuming. Therefore, setting up a false trail is a good delaying tactic that is easy to do and does not require much time. When moving along a stream and using it as a deception technique, the fact that you are in the stream will slow down anyone in pursuit. Even greater success can be achieved by entering and leaving the stream carefully. (See Figures 1-20 and 1-21.) Some of the following points will also aid in eluding a following party:

- Stay in the stream for 100 to 200 meters.

- Keep in the center of the stream and deep water.

1-36

Figure 1-19. False trail leaving stream.

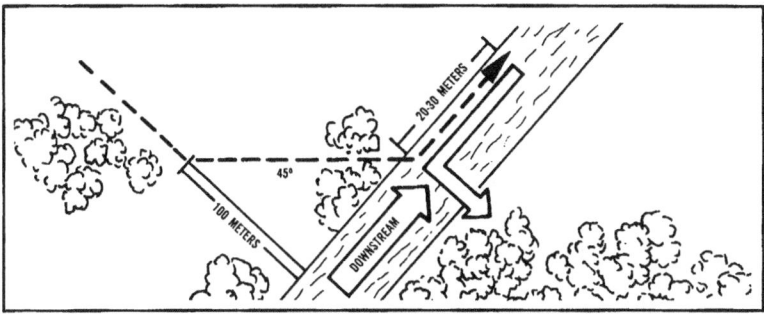

Figure 1-20. Crossing stream.

Figure 1-21. Parallel trail.

1-37

- Watch for rocks or roots near the banks that are not covered with moss or vegetation and leave the stream at this point.
- Walk out backward on soft ground.
- Walk up small, vegetation-covered tributaries and replace the vegetation in its natural position.
- Walk downstream until coming to the main river, then depart on a log or pre-positioned boat.
- Enter the stream, having first carried out the above tactic, then exit at the point of entry and make a large backward loop, crossing and checking it, and move off in a different direction.
- Using a stream as a deception technique is one of the best ways to slow down and lose a following party. The deception starts 100 meters from the stream and the successful completion of the tactic is to ensure that the following party does not know where to exit from the stream.

An experienced tracker can usually cover his tracks because he knows what the natural vegetation of the area looks like and he can replace it in a truly natural state. However, unless you are that well trained, don't try it.

When being tracked by trained, persistent enemy trackers (those you know are there either by hearing them or seeing them), the best bet is to try to outrun or outdistance them or double back and ambush them, depending on their strength and yours.

Camouflage Techniques Used to Confuse Enemy Trackers

Camouflage techniques used in countertracking are the same as those used in tracking. Camouflage is discussed on page 1-24.

Techniques Used to Confuse Dogs

At times enemy tracking teams may employ dogs; some techniques that may be used are to—

- Scatter black or red pepper or, if authorized, a riot control agent, such as ortho-chlorobenzalmalononitrile (CS) powder along the route used.
- Employ silence suppressed weapons against unleashed animals.

SELECTION OF DEFENSIVE MEASURES

In the process of selecting defensive tactics, SF and guerrilla leaders must fully consider and evaluate available measures. While a total list of defensive measures is endless, measures to be considered should include—

- Having auxiliaries and underground increase counterintelligence activities.
- Initiating diversionary activities in other areas.
- Intensifying operations against lines of communication.
- Preparing to implement guerrilla base evacuation plan.
- Instituting delay and harassing tactics.
- Exploiting the guerrillas' inherent advantages of fluidity and intimate knowledge of terrain.

1-38

- Preparing to initiate *breakout* operations (see paragraph below).
- Withdrawing to more favorable terrain.
- Increasing frequency of ambush operations.
- Preparing for the enemy's use of chemical and biological weapons.
- Establishing caches in potential withdrawal areas.
- Emphasizing passive air defense measures.
- Planning for employment of concentrated fires of automatic and semi-automatic weapons against helicopters and low-performance aircraft.
- Planning counteractions against enemy heliborne and airborne operations.
- Taking adequate communication security measures.
- As a last resort, implementing dispersal plan. This plan must include instructions covering interim conduct and ultimate reassembly.

BREAKOUT OPERATIONS

Acting on Indications of an Encirclement

As discussed earlier, encirclement by counterguerrilla forces poses the greatest threat to guerrilla forces. The SF and guerrilla commanders must be constantly on the alert for indications of an encirclement. When they receive indications that an encircling movement is in progress, the guerrillas immediately maneuver to escape while enemy lines are still thin and spread out and coordination between advancing units is not yet well established. If escape is not accomplished and the enemy completes its encirclement, the guerrilla force attempts a breakout.

Planning a Breakout

If an encirclement is a difficult operation, a breakout from encirclement is equally difficult. Unless the encircled guerrilla force has explicit orders to defend in place, the force should break out and execute the operation before the enemy is able to establish an organized containment. The need for quick decision making, however, should not lead to an attempted breakout without adequate planning. Planning considerations involve eight factors.

Area for the attack. The guerrilla force should launch the main attack against enemy weakness in a direction that will ensure breakthrough in the shortest possible time. Commanders should indicate the direction of attack by designating objectives and an axis of advance. They should assign objectives that ensure penetration of the encircling force and preservation of the gap created.

Time of attack. Since deception and secrecy are essential to a successful breakout, the SF and guerrilla commanders may decide to attack during darkness or other periods of limited visibility. They must also consider effectiveness of enemy air in selecting the time for the breakout. When the enemy has local air superiority, the guerrilla force may have to conduct the breakout at night or when weather conditions reduce the enemy's effectiveness.

Organization. An encircled guerrilla force is usually organized into four distinct tactical groups for the breakout.

The *rupture force* may vary in size from one-third to two-thirds of the total encircled force. Its mission is to penetrate the enemy encirclement and widen the gap. The rupture force also holds the shoulders of the gap until all other encircled forces can move through. After all other encircled forces have passed through the penetrated area, the rupture force may be employed as a rear guard.

The *reserve force* provides rear and flank security and may assist the rupture force by conducting diversionary attacks. It may become the advance guard for further movement when freedom of action is gained.

The *rear guard* provides rear and flank security. It may also conduct diversionary attacks to assist the rupture force. When freedom of action is gained, this force becomes the advance guard for further movement.

The *main body* consists of all guerrilla forces not assigned to one of the other three elements.

Deception. Effective deception may be achieved by employing feints, diversionary attacks, and demonstrations. These measures are designed to deceive the enemy as to the location of the main attack. If sufficient guerrilla forces are available, these deception measures may allow the guerrillas to break out at more than one point.

Concentration of forces. Prior to the breakout, the guerrilla force must gradually change from defending the perimeter to forming a strong breakout force. As the situation permits, every element that can be spared from the perimeter must assemble for employment in the breakout.

Communication. Secrecy is essential to the success of breakout operations. Messengers should therefore be used within the encircled unit. Radio and wire may be used, but transmissions must be closely guarded.

Logistics. The SF and guerrilla commander must make plans to relieve the guerrillas of all equipment and supplies not essential for fighting during the breakout. Nonessential equipment and supplies are destroyed or cached.

Execution. Once the plans are completed, the guerrillas execute the breakout. Since secrecy and security are essential, the operation must follow a strict sequence of events:

- Guerrilla scouts locate a weak point in the enemy's line of encirclement. The location should be along an axis of movement that will benefit the guerrillas following the breakout.

- Elements on the defensive perimeter participating in the breakout are released from their defense mission. These elements are assembled with their respective tactical groups as late as possible before the breakout begins.

- At the scheduled time of attack, the rupture force, supported by indirect fires (and tactical air when available) effects the penetration, widens the gap, and holds the shoulders of penetration.

- The main body and reserve force then pass through the gap and continue the attack to the assigned objective.

- The rear guard withdraws on order and follows the reserve force through the gap.

- When all encircled forces have passed through the gap, the rupture force withdraws, prepared to fight a rear guard action.

- The guerrillas break contact with the enemy as rapidly as possible. They proceed to MSSs or to alternate bases to prepare for future operations.

CHAPTER 2

RECONNAISSANCE, SURVEILLANCE, AND TARGET ACQUISITION

Special Forces teams operate deep behind enemy lines and are ideally situated to contribute to the overall intelligence collection effort. Special Forces reconnaissance patrols support the reconnaissance, surveillance, and target acquisition (RSTA) effort by gathering and reporting information or by emplacing sensor devices to be monitored by higher headquarters. This chapter describes the types of reconnaissance and surveillance patrols and discusses RSTA organization, equipment, and countermeasures.

Section I. The Reconnaissance Patrol

Reconnaissance patrols provide the commander with timely, and accurate information of the enemy and the terrain it controls. This information is vital in making tactical decisions.

MISSIONS

The commander asks questions about the enemy and about the terrain. These questions are missions for reconnaissance patrols. For example, reconnaissance patrols can provide answers to—

- Questions about the enemy. Where is the enemy located? What is its strength at a certain location? How is it equipped? What is it doing?

- Questions about the terrain. How deep are the streams? How wide? Are the banks too steep for armored vehicles? Are bridges and roads in the area mined or damaged?

These questions are answered when the missions are accomplished.

TYPES OF RECONNAISSANCE AND SURVEILLANCE

Depending on the informational needs of the commander, three types of reconnaissance can be performed.

Area Reconnaissance

The commander may require information on a specific location or a specific area, usually a known or suspected position or activity. An area reconnaissance patrol secures this information by reconnoitering the location or area or by maintaining surveillance over the location.

2-1

Zone Reconnaissance

The commander may require information on an extended area, or he may desire information on several locations within a zone. A zone reconnaissance patrol gets this information by reconnoitering the zone, by maintaining surveillance over the zone, or by coordinating area reconnaissance of designated locations within the zone.

Route Reconnaissance

The commander may require information about the enemy, obstacles, route conditions, and critical terrain along a specific route.

SIZE

The minimum size of a reconnaissance patrol is two men. There is no maximum. Size (above the minimum) is determined by the mission.

TYPICAL METHODS OF ORGANIZATION

If the objective to be reconnoitered is restricted in size and is clearly defined (as in most area reconnaissance), the security element can normally perform its mission from one location. Figure 2-1 is one type of organization.

If, as in most zone reconnaissance, the objective is not clearly defined or located, and movement into or through the objective area by mutually supporting bounds is necessary, the reconnaissance patrol is effectively configured as shown in Figure 2-2.

In a small reconnaissance patrol, the patrol headquarters may form a part of the reconnaissance element or one of the R & S teams, rather than being a separate element. The number and size of the various teams and elements must be determined through the analysis of mission, enemy, terrain, troops, and time available (METT-T).

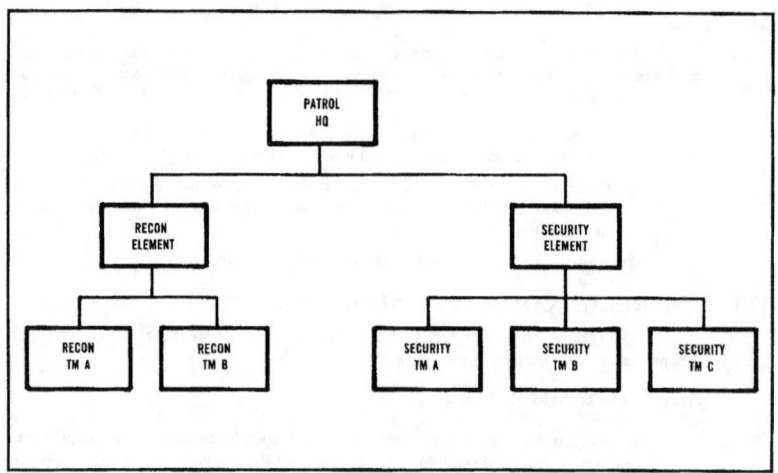

Figure 2-1. Typical organization

2-2

PLANNING

The patrol leader plans the actions to be taken in the objective area, based on the information he gains from maps, photos, previous reconnaissance reports, and other intelligence. The leader's plan for actions at the objective includes establishment of the objective rallying point (ORP), conduct of the leader's reconnaissance, the patrol's reconnaissance of the objective, withdrawal, and dissemination of information. Establishment of the ORP, withdrawal, and dissemination of information are discussed on page 2-11 of this chapter.

FUNDAMENTALS

There are three fundamentals that apply to all types of reconnaissance operations. These fundamentals provide a basis for planning, developing techniques, and executing successful reconnaissance.

Gain All Required Information

The parent unit must tell the patrol leader what specific information is required. A patrol's mission may be specific as to what information is required. Specific information requirements (SIR) are included in the patrol's operation order. Special information requirements may include information to be gathered en route and at the objective. During the patrol, members must continuously exchange all information acquired. The patrol reports all the information it has gathered, but its mission is not complete until all required information has been gained.

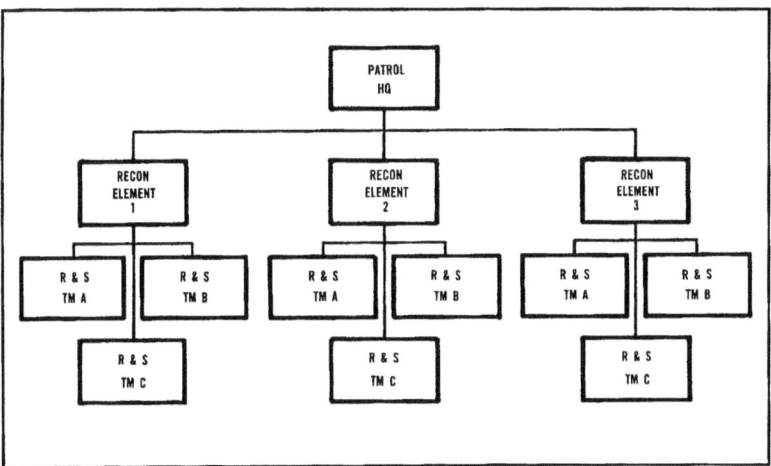

Figure 2-2. Patrol configuration.

Avoid Detection by the Enemy

A patrol must not let the enemy know it is in the objective area. If the enemy knows it has been observed, it may move its elements or change its plans. If the enemy detects a patrol, it will increase its security measures to keep the patrol from completing its mission. The patrol should minimize its movement in the objective area. It should move no closer to the objective than necessary. If possible, it should use long-range surveillance or night observation devices to gain information. Camouflage, discipline, and stealth help to avoid detection. The patrol should plan routes that limit the effectiveness of enemy radar and RSTA devices. Also, it should minimize radio traffic to avoid enemy radio direction finding (RDF).

Employ Security Measures

If detected, a patrol must be able to break contact and return to friendly lines with the information it has gathered or continue the mission, if possible. The patrol should rehearse plans for breaking contact, including handling casualties. It should have security teams with machine guns in overwatch positions with good observation and fields of fire. Security teams should be able to hit the enemy with direct and indirect fire to help the reconnaissance teams break contact. Security teams and sensor devices may be used to give early warning of the approach of enemy troops.

AREA RECONNAISSANCE

An area reconnaissance patrol reconnoiters a specific objective and its immediate surroundings (Figure 2-3). The objective's location is usually stated in grid coordinates: "EL987453," or as a terrain feature: "the junction of Highway 9 and this road." An area reconnaissance may be a primary mission or it may be a part of a zone or route reconnaissance. For example, a reconnaissance of a bridge site could be an essential part of a route reconnaissance. A patrol can use long-range observation or surveillance or short-range observation or surveillance. It can adapt or modify according to the situation. It may develop other methods as long as the basics of reconnaissance patrolling are applied. Single or multiple R & S teams may be used with either method. Security measures are based on the situation.

Long-Range Observation or Surveillance

Long-range observation or surveillance is the act of watching an objective from a point (an OP) that is far enough from the objective so as to be outside enemy small-arms range and its local security measures. This method should be used whenever METT-T permits the required information to be gathered from a distance. It is the most desirable method for executing an area reconnaissance, since the patrol does not get close enough to the objective that it risks being detected. Also, if the patrol is discovered, artillery and other fire can be employed on the objective without endangering the patrol.

When all required information cannot be gathered from one direction, a series of OPs may be occupied by one or more reconnaissance patrols. The patrol should use OPs that have cover and concealment. They should permit good view of the objective. Routes between OPs and from OPs to the ORP should have cover and concealment.

Security teams and sensor devices are used to give warning. Security teams are positioned where they can help reconnaissance teams break contact by shooting and by calling for indirect fire.

2-4

Figure 2-3. Area reconnaissance patrol reconnoiters a specific objective.

2-5

Short-Range Observation or Surveillance

Short-range observation or surveillance is the act of watching an objective from a place that is within the range of enemy local security measures and small-arms range.

Reconnaissance teams use short-range observation when METT-T requires close approach to the objective to gain the required information.

Short-range observation or surveillance may be from a preselected OP, but usually the reconnaissance teams must move near the objective before they can find a position from which to observe. In some cases they may be able to gather the required information by listening without visual observation.

Short-range observation increases the chance the patrol will be detected. The enemy employs anti-intrusion devices close to its key installations. The reconnaissance teams must frequently pass through OPs, defensive wire, and minefields to get close enough to gain the required information. They use inclement weather to cover the sounds of their movement. Reduced visibility favors short-range observation.

When short-range observation is necessary, the reconnaissance teams use every measure possible (both passive and active) to avoid detection. A security element is less effective on a short-range observation because the reconnaissance teams may mask supporting fire, should the enemy attack the patrol. To preclude noise and reduce the chance of detection, the patrol leader may make the reconnaissance teams as small as 2-man teams. The rest of the patrol can be assigned security tasks.

Leader's Reconnaissance

In most cases, the size of a zone or route reconnaissance objective precludes the timely completion of a leader's reconnaissance. However, area reconnaissance objectives within the zone require a leader's reconnaissance.

The purpose of a leader's reconnaissance for a reconnaissance patrol is to determine whether the plan for actions at the objective needs to be modified and to ensure a smooth execution of the reconnaissance. All the tasks of a leader's reconnaissance are not performed for every objective. The patrol leader should not permit the leader's reconnaissance to compromise the patrol. He may restrict the scope of the leader's reconnaissance based on the situation.

The leader's area reconnaissance of an objective may have the following tasks:

- Pinpoint the objective. If possible, this is done by checking terrain features in the area and not by directly approaching the objective.
- Locate observation or surveillance positions, routes, and security positions the patrol will use.
- Determine the enemy situation in the objective area, locate enemy OPs, determine enemy alert status and activity, and accustom the patrol to the local sounds in the area.
- Designate the release point or positions that elements will occupy.

ZONE RECONNAISSANCE

A zone reconnaissance is conducted to obtain information on enemy terrain and routes within a specific zone. The zone is specifically defined by boundaries.

A patrol leader may be directed to conduct one or more area reconnaissances within his zone, or he may decide that such area reconnaissances are necessary based on analysis of the mission and SIR.

For most reconnaissance patrols, separate security teams are impractical because they cannot observe enough of the area to be effective. Normally, R & S teams are used. Each R & S team provides its own security while making a reconnaissance of its sector or part of the zone.

A zone reconnaissance may be conducted by using single or multiple reconnaissance elements. As in an area reconnaissance, the following techniques may be developed as long as the fundamentals to reconnaissance patrolling are applied.

A single reconnaissance element is favored when—

- Mission information requirements and SIR can be gathered within the required time by a single reconnaissance element.

- Control of multiple elements in the objective area is difficult.

- Terrain is open and visibility is good.

- Enemy security measures such as patrols, sensors, and radar are active in the area.

Multiple reconnaissance elements are favored when—

- A patrol may have to reconnoiter a large area in a relatively short time. In this case, it must employ multiple reconnaissance elements to finish on time and to gain the required information.

- The patrol leader may control the reconnaissance elements from one location, or he may lead one of the elements himself. He may give one or more point reconnaissance missions to each R & S team. His instruction to the reconnaissance elements' leaders may be general or specific, but he must at least state mission information requirements and SIR. His instruction usually tells the route they will follow in the form of a direction and distance.

Methods for Multiple Reconnaissance Elements

There are three methods for the multiple reconnaissance elements to use. They must not set a pattern by habitually using the same method of covering an area.

Converging routes method. The reconnaissance elements may move through the area on converging routes (Figure 2-4). Beginning at an ORP, where elements are briefed, they move on separate routes through the area and converge at the end of the area at a rendezvous point. The patrol leader briefs each reconnaissance element on the route it is to take, the location of the rendezvous point, and the linkup time at the rendezvous point.

The converging routes method may be repeated several times. It requires skill in land navigation to find the rendezvous point. The rendezvous point should be an easily identifiable terrain feature. Time may be wasted when elements completing their routes early must wait on other elements before continuing.

Fan method. There are variations of this method (Figure 2-5). Reconnaissance and security teams leave from and return to the same ORP. Therefore, less skill in land navigation is required because they are returning to a place they have been before. Teams

2-7

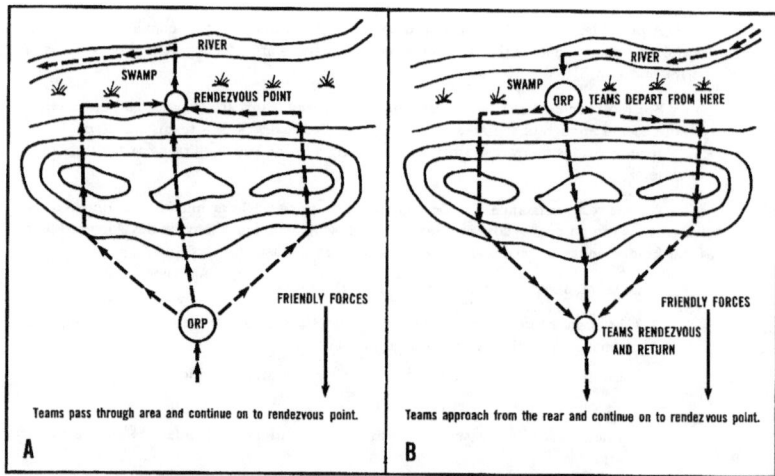

Figure 2-4. Converging routes methods of reconnaissance.

may reconnoiter their routes singularly or simultaneously. Routes are planned so that reconnaissance elements do not converge and interfere with each other.

This method tends to compromise the location of the ORP after a time because of so much movement into and out of the ORP. The ORP should be relocated occasionally. Also, reconnaissance elements must make sure they are not followed back to the ORP by enemy trackers.

Successive sector method. The patrol leader may divide the objective area into segments and assign each reconnaissance element a segment (Figure 2-6). As each element moves at its own pace until it has completed its part of the reconnaissance, no time is wasted in waiting. It also allows reconnaissance element leaders more freedom of action. The patrol leader may only dictate the general direction of movement the reconnaissance elements will take in their area of operation (AO), the time for completion, and the linkup or rallying point.

ROUTE RECONNAISSANCE

A route reconnaissance is the survey of a route to get information about the enemy, obstacles, route conditions, and critical terrain along that route.

Planning a route reconnaissance is as detailed as planning a zone reconnaissance. The patrol leader identifies places along the route where area reconnaissance is necessary, for example, bridges, obstacles, and road junctions. He may have single or multiple reconnaissance elements. When an area reconnaissance is required, the patrol leader may conduct it using long- or short-range observation or surveillance, as previously described.

A commander may order a route reconnaissance when he needs information on routes to his objective, or to alternate or supplementary defense positions. Usually, a patrol is given an overlay of the route it must reconnoiter. A patrol can reconnoiter a route without

2-8

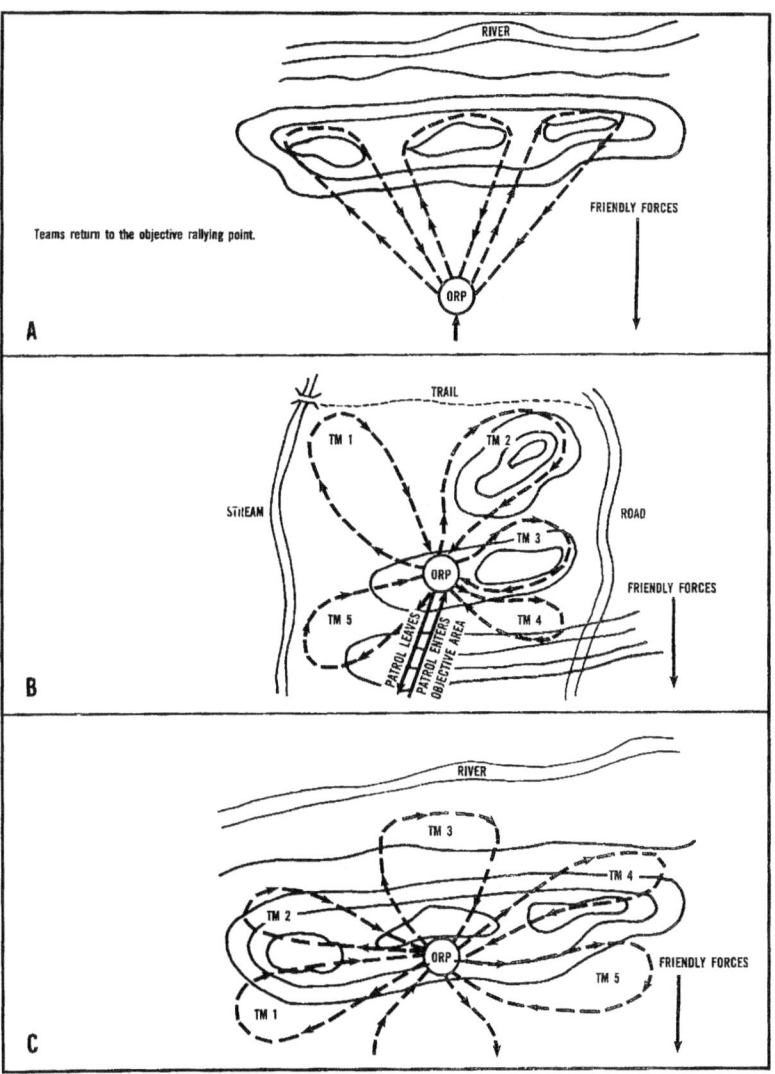

Figure 2-5. Fan methods of reconnaissance.

2-9

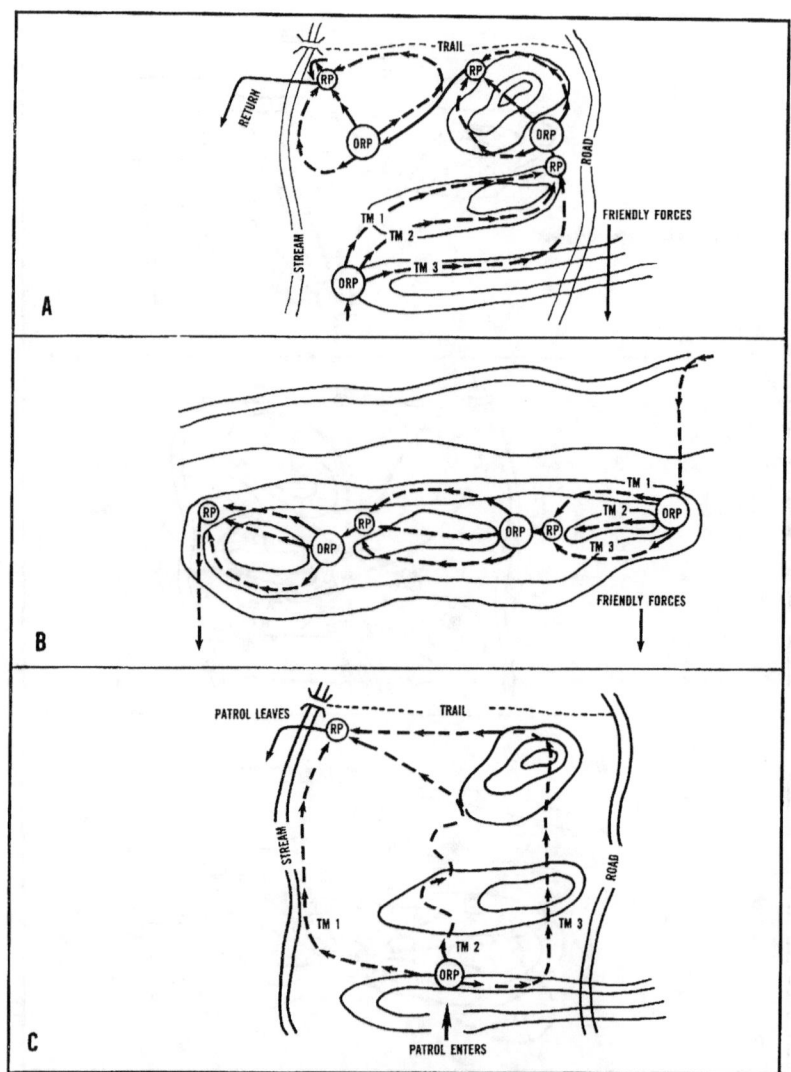

Figure 2–6. Successive sector methods of reconnaissance.

2–10

the attachment of specially qualified troops. The patrol is given the SIR just as for any other reconnaissance mission. The SIRs for route reconnaissance patrols are—

- The available space in which a unit can maneuver without being forced to bunch up because of obstacles (reported in meters). The size of trees and the density of forests are reported because of their effect on vehicular movement.

- The location and type of all obstacles and the location of any available bypass. Obstacles can consist of minefields, barriers, steep ravines, marshy areas, or nuclear, biological, and chemical (NBC) contamination.

- Enemy forces or positions along the route. The patrol uses the SALUTE (size, activity, location, unit or uniform, time, equipment) report format.

- Observation and fire both from and into the route. This information will greatly assist planners as a supplement to map information.

- Locations along the route that provide good cover and concealment, especially from the air.

Deliberate route reconnaissance is discussed in detail in FM 5-36. The technical nature of the data necessary for route classification requires attachment to the patrol of intelligence or engineer troops who are qualified to classify a route. An alternative to attaching specialists is to give the patrol detailed SIRs that require it to gather the specific technical data required for classification. In this case, the patrol will report the data to qualified analysts in the rear area who then classify the route.

WITHDRAWAL FROM AN OBJECTIVE

Once the mission on an objective has been accomplished, the patrol withdraws quickly to the ORP or a rendezvous point and reorganizes. A withdrawal to the ORP is similar to a withdrawal to a preselected rallying point.

Reconnaissance patrols make sketches and take photos of the objective. Additional copies are made to ensure information is returned.

Reconnaissance patrols withdraw from the objective when they have collected all the information possible. Security teams remain in position until the rest of the patrol has departed the objective area. Teams returning to the ORP reoccupy positions and reestablish security.

DISSEMINATION OF INFORMATION

Once the security teams have returned, they prepare a status report. Element leaders account to the patrol leader for all men and equipment, and tell him the status of weapons and equipment.

If the patrol has not been detected, the patrol leader disseminates information at the ORP from a safe distance (normally one terrain feature away), or during movement.

Element leaders debrief their men and move to the center of the perimeter to give the information to the recorders. Element leaders must be allowed time to disseminate the information to their men.

Recorders write down the information as the element leaders give it to them.

The recorders consolidate the information and read it back to the element leaders.

2-11

Section II. RSTA Equipment and Countermeasures

An SF reconnaissance patrol may use RSTA equipment in the field to maintain security, to help it move to or from the objective area, and to accomplish its mission at the objective. Since the enemy has an equal or superior RSTA capability, the reconnaissance patrol must take measures to minimize the effectiveness of the enemy's RSTA devices and to counter its RSTA effort.

EQUIPMENT

There are four categories of RSTA equipment: night observation devices, ground surveillance radars, remote sensors, and airborne sensors. Currently there is no firm organization or basis for some of this equipment. Some items are in the final stages of development and will soon be available. For specific items of equipment, see Student Handbook (SH) 7–285. RSTA equipment and organization varies at each echelon depending on the type unit; for example, armor, light infantry, mechanized infantry, or special operations.

Night Vision Devices

Night vision devices have their own light source and can illuminate a close–in object. Active infared (IR) devices, such as the metascope, may be used in a passive role although they have an active capability. Infared devices are easily detected by the enemy, they are vulnerable to high–intensity light, and precipitation degrades their efficiency.

Passive night vision devices do not emit any light or radiation of their own. Passive devices gather light that is reflected from or emitted by the target. This may be natural light from moonlight or starlight (ambient light) or it may be artificial light from some other source (for example, IR, flares, searchlight). Passive devices become reflective when the available ambient light drops a certain level.

Night vision devices may magnify their targets, thus, increasing their range and usefulness. Most US–made night vision devices have an image intensification capability.

A wide range of night vision devices are available. They are lightweight and compact enough to be carried and used by patrols. These night vision devices include the starlight scope, thermal imagery devices, electronic binoculars (night vision goggles), and metascopes. In addition, there are other types of night vision devices that are too heavy and bulky to be carried by patrols but may be used in support of patrols. These include searchlights and night observation devices (NOD). Searchlights can be used to provide IR or white light to illuminate an area for a patrol using passive night vision devices. They may also be used to temporarily blind enemy troops using night vision devices by subjecting them to a direct flow of high–intensity IR or white light.

Remotely Employed Sensors (REMS)

Patrols may use sensors to augment their security forces in both combat and reconnaissance missions. Sensors may be used to monitor trails that lead into the objective area when adequate troops are not available to establish an OP on each trail or road. Sensors can also be used to give early warning to a patrol occupying a patrol base.

A patrol may have a mission to install one or more hand–emplaced REMS behind enemy lines. The patrol leader plans and organizes similar to the way he plans and organizes for a mine emplacement mission. Selected men must be trained and briefed on the devices they will be emplacing. Specialists on the devices may be attached to the patrol.

The patrol leader will make use of the REMS that are already deployed in his area of operation by coordinating with the S2 or G2 to get reports of sensor-derived intelligence near the objective and along the patrol route. These REMS may have been emplaced by artillery, helicopters, high-performance aircraft, or by other patrols.

Patrol members should know the five basic types of REMS: seismic, magnetic, acoustic, disturbance, and active IR. In most cases, sensors are emplaced in clusters like mines, and each cluster will contain at least two different types of sensors.

Seismic REMS detect the vibrations in the ground caused by a vehicle or man crossing the terrain nearby. The detection of the seismic sensor is limited by the ability of the soil to transmit vibrations. Seismic sensors must usually be emplaced in large numbers or in a detailed surveillance plan to be effective.

Magnetic REMS detect the movement of ferrous metal through a magnetic field. They are usually used in conjunction with other sensing devices. They will not recognize intruders who do not carry ferrous metals.

Acoustic REMS detect audio frequencies by means of a microphone. They can use spectrum analysis to give a better definition of the target source. Acoustic REMS are most effective in the commandable mode. When the intruder is detected by another source, the acoustic sensor is activated to permit identification. Acoustic sensors will pick up all sounds on the battlefield and an enemy can use them to send out erroneous information as part of a deception plan.

Disturbance REMS react to physical contact with the target. To be effective, an area has to be seeded densely with this type sensor. The intruder must have physical contact and there is a high false alarm rate with this type sensor. It is very useful for monitoring movement into a new area or for locating targets for indirect or aerial fire.

Active IR REMS employ the principle of line-of-sight which detects an interrupted beam of light. The IR sensor must be carefully employed so the beam has nothing to interrupt its path. It has a high false alarm rate. It is used on roads, trails, or perimeters to detect movement in that location.

Radar is one of the few RSTA items that almost has an all-weather, day and night capability. Of all the surveillance equipment now in use, radar has the greatest range for detection of moving targets. Radar does depend on line-of-sight detection and its ability to identify targets is relatively limited. At present, no radar is available that is compact and lightweight enough to carry on a patrol. A wide variety of ground surveillance radars are available which can support a patrol from a friendly position. Ground surveillance radars may be used to assist the patrol in navigation by vectoring the patrol to a friendly position or through an enemy position. Airborne radars are used to develop intelligence that may be used in planning a patrol.

RSTA COUNTERMEASURES

Threat forces have a RSTA capability similar to that of US forces. The enemy will use its available RSTA assets to find patrols, to secure its key installations, to provide early warning of patrol movement, and to conduct deception operations to counter US RSTA operations. Patrols should be alert for enemy deception operations when making reports on enemy operations.

In most cases, the enemy makes maximum use of its RSTA assets at the most critical parts of the battle area, such as along the main line of defense (MLD). It also uses RSTA batteries and communications installations.

The patrol leader should select heavily vegetated routes to reduce the effectiveness of enemy radar and night vision devices. When a patrol must move across flat, open areas, such as water or certain types of desert, the patrol should move across the area as quickly as possible and plan for electronic countermeasures or deception operations to counter the enemy radar threat. Ground surveillance radar is less effective against small groups of men. In an area where the enemy is using radar, a patrol may infiltrate in groups of two or three men and link up behind the enemy MLD in a secure, heavily vegetated area to make final preparations for actions at the objective. Darkness provides no concealment from an enemy using night vision devices. The patrol maintains dispersion between men even at night.

Weather reduces the effectiveness of enemy RSTA equipment. Rain, snow, fog, and other inclement weather can be used to assist in concealing patrol operations from an enemy force that is using RSTA equipment. Other precautions should not be neglected, however, since precipitation may not completely screen the patrol from enemy observation and the weather is subject to change.

CHAPTER 3

LAND NAVIGATION

Movement on land must be carefully planned. The patrol leader is responsible for getting the patrol to and from the objective safely. To do this, he must maintain his orientation on the ground. Normally, the starting location and destination are known if a map is available. But maps are often unavailable for a great portion of the globe; hence, the possibility exists that patrol may have to move overland without a map or without the time to make one. This chapter describes several improvised methods of land navigation during daytime and nighttime movement.

GENERAL AZIMUTH METHOD

For this method, use a means other than a straight line azimuth for maintaining direction of movement. The patrol leader may pick a terrain feature, such as a ridge, stream, or the edge of a body of water to use as a guide during movement. However, he must remain oriented on the map and check the checklist.

Advantages. The advantages of this method are that it speeds up movement, avoids fatigue, and often simplifies navigation as the terrain feature followed is a constant checkpoint.

Disadvantages. Following a terrain feature can be dangerous as it usually puts the patrol on a natural line of drift. This is especially true between friendly and enemy lines or any place where the enemy has tight security.

Techniques for Movement Along Terrain Features

Ridges and valleys. Routes are shown for units moving downhill along a ridge and valley (Figure 3-1). General azimuths are shown for portions of each route. As some hilltops are round, finding the beginnings of ridges and valleys may be difficult. If so, the patrol should go to the top of the hill and proceed on the desired general azimuth until it is established on the correct terrain feature. If a patrol following the ridge notes that the ridge is on a general azimuth of south within 200 meters of the hilltop, or that the ridge is on a general azimuth of southeast anywhere along the route, it knows that it has taken the wrong ridge. But, by using its pace count and by determining which ridge it is on by terrain association, the patrol can still find its location and either continue on a new route or move to the route planned first and continue. At night, routes are memorized to avoid map checks which are time-consuming and risky if a light is used. If PVS-5 night vision goggles are available, they can be used for map checks.

Rivers and streams. The starting point must be accurately determined. See Figure 3-2. Proper use of the pace count will avoid errors such as turning at the wrong stream junction or a junction with a stream not shown on the map. Direction of stream flow serves as an additional check. The stream flows generally north along the first portions of the route.

3-1

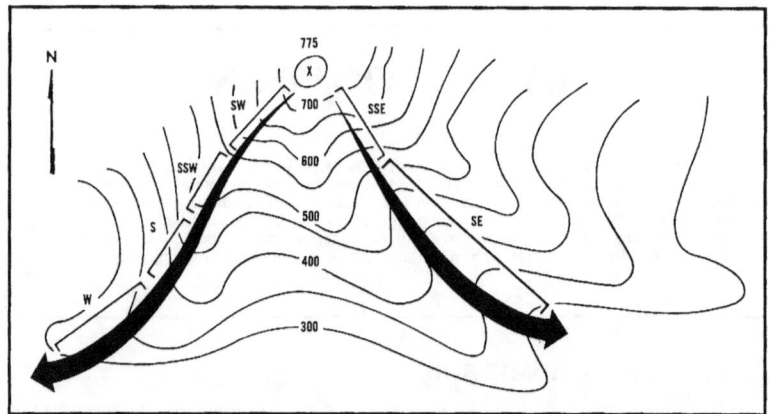

Figure 3-1. Movement along ridges and valleys.

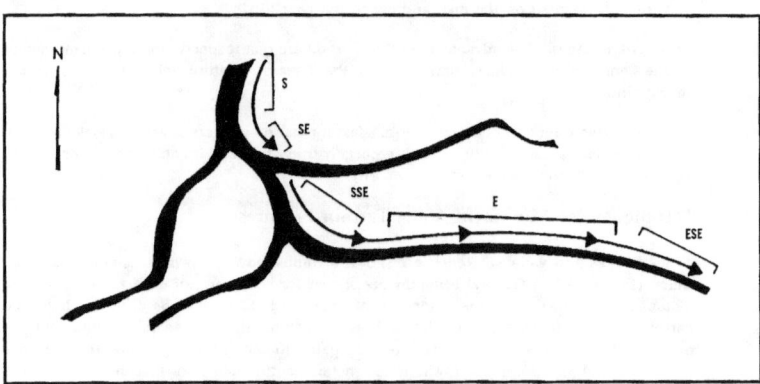

Figure 3-2. Movement along rivers and streams.

Use by Mounted and Airmobile Patrols

Mounted patrols can use similar techniques on charted road nets. Where vehicle movement is not limited to roads, use the same techniques as foot troops. Use the vehicle odometer instead of pace count.

Airmobile patrols flying low level or nap of the earth can use similar techniques, but must use visible checkpoints and time of flight instead of pace count.

3–2

DEAD-RECKONING METHOD

Use dead reckoning to aid navigation when there are no recognizable terrain features (as in swamps and certain deserts) or when features cannot be seen (as in heavy forest). Use dead reckoning to move from one checkpoint to another or for an entire movement. (Use checkpoints when they are available.) Dead reckoning has three parts: an azimuth, a distance (meters), and a starting point.

The dead-reckoning method only requires a compass and a means of measuring distance such as a pace man. A map is recommended, however, for military tasks such as calling artillery or confirming terrain.

Make sure that your location (starting point) is pinpointed exactly. You may have to reconnoiter to determine this.

Record the distance traveled. Have at least one pace man who knows how many of his paces equal 100 meters. He should walk near the patrol leader so he can easily tell him the pace count (in hundreds of meters), for example, "Pace count, 500 meters." The pace man keeps track of the pace count by using a knotted string or some other device. A technique is to use two pace men and average their distances.

Control the direction traveled. The lead patrol has a compass man, and the patrol leader is told the azimuth to follow. The main body of the patrol also has a compass man who follows the same azimuth as the compass man in the point team. The patrol leader must keep tight control of the direction of movement as a slight deviation can cause big problems over an extended distance. However, the patrol should avoid danger areas by using a detour or bypass.

Make maximum use of checkpoints. Checkpoints can be linear terrain features such as a road or a stream, or other features such as prominent hills. You can use two artillery rounds fired in such a way that the patrol's position can be verified by resection.

At every checkpoint compare the dead-reckoning location with the patrol's actual location in reference to the checkpoint. For example, a patrol has been using dead reckoning for 2,000 meters on the first leg of its route. A trail is intersected. The patrol leader checks the direction of the trail and its contour against the map to see if the patrol intersected it where planned. If not, he adjusts the route based on the known location.

FINDING DIRECTION BY DAY

If you do not have a compass, you can use the sun to find approximate true north (and from north, any other direction). The method explained in Figure 3-3 can be used any time the sun is bright enough for a stick placed in the ground to cast a shadow. To determine direction by means of a watch, follow the steps in Figure 3-4.

FINDING DIRECTION BY NIGHT

Usually you must move more quietly at night than in day. Here are some general rules to help you. Move around thick undergrowth, dense woods, and ravines. Your field of observation is reduced and it is difficult to move quietly. Move as quickly as circumstances allow, but avoid running if possible. You may fall or make unnecessary noise. If you do not have a compass, you can find direction by other methods.

3-3

71

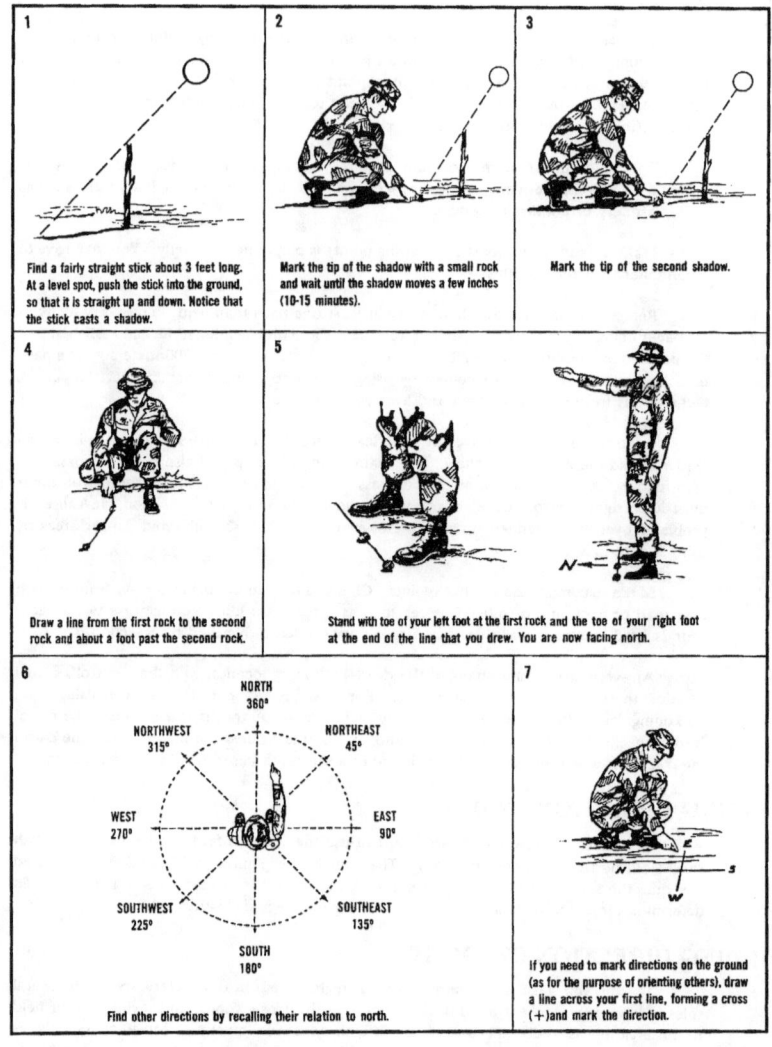

Figure 3-3. Finding direction by the sun.

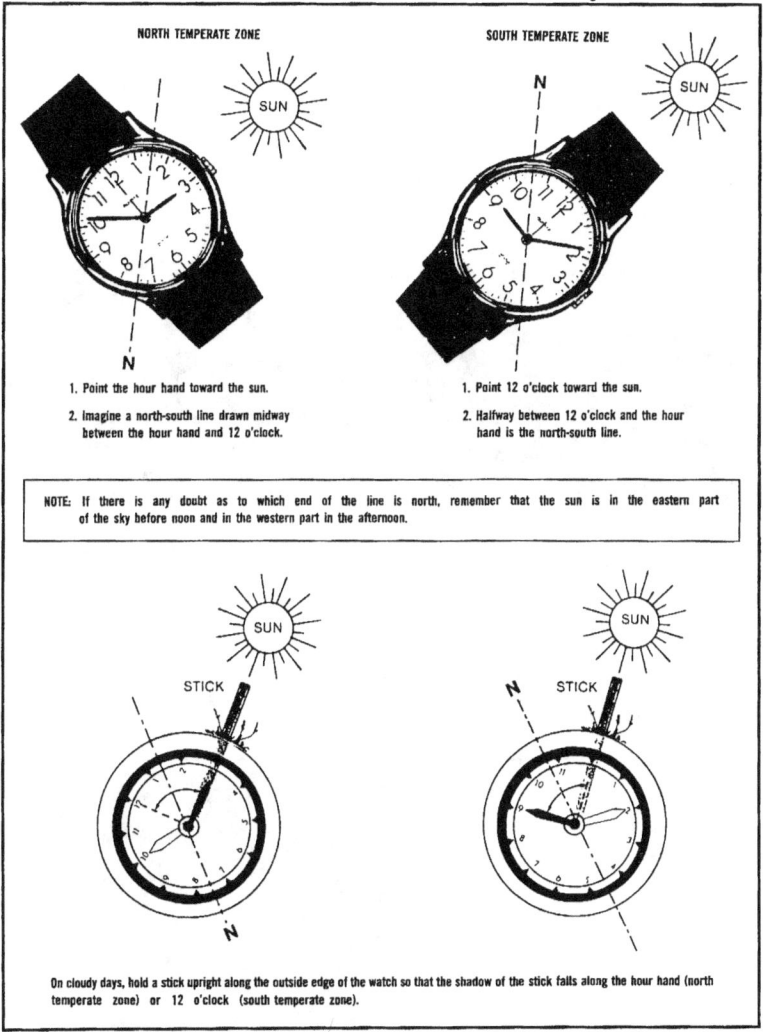

Figure 3-4. Using a watch to find north.

The North Star

North of the equator, the North Star shows you true north. To find the North Star, look for the Big Dipper or the Cassiopeia constellation (Figure 3-5). The Big Dipper rotates slowly around the North Star and does not always appear in the same position. The constellation Cassiopeia has five bright stars shaped like a lopsided M (or W, when it is low in the sky). Cassiopeia also rotates slowly around the North Star and is always almost directly opposite the Big Dipper. Its position, opposite the Big Dipper, makes it a valuable aid when the Big Dipper is low in the sky, possibly out of sight because of vegetation or high terrain features.

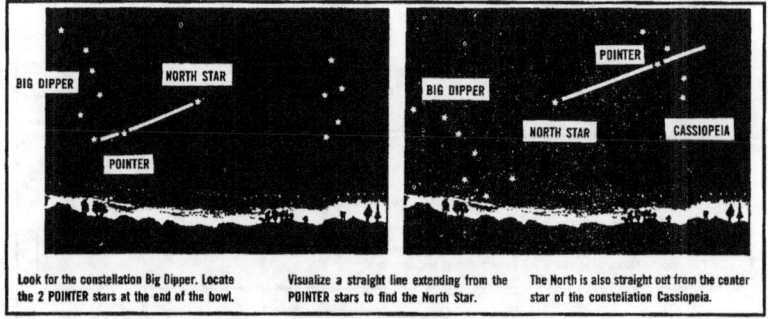

Look for the constellation Big Dipper. Locate the 2 POINTER stars at the end of the bowl.

Visualize a straight line extending from the POINTER stars to find the North Star.

The North is also straight out from the center star of the constellation Cassiopeia.

Figure 3-5. Locating the North Star.

The Southern Cross

South of the equator, the constellation Southern Cross helps you to locate the general direction of south, and from south, any other direction. This group of four bright stars is shaped like a cross that is tilted to one side (Figure 3-6).

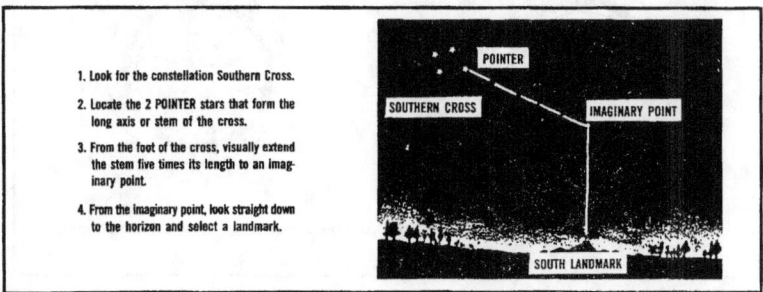

1. Look for the constellation Southern Cross.

2. Locate the 2 POINTER stars that form the long axis or stem of the cross.

3. From the foot of the cross, visually extend the stem five times its length to an imaginary point.

4. From the imaginary point, look straight down to the horizon and select a landmark.

Figure 3-6. Locating south by the Southern Cross.

3-6

CHAPTER 4

GROUND MOVEMENT
AND MANEUVER

This chapter deals with the movement of an element from infiltration through movement within the area of operations and includes stay behind operations. It is divided into two major sections: *foot movement* and *stay behind*. Foot movement is of primary importance because an element must plan for the method even though it may travel most of the distance by some other means.

Section I. Foot Movement

Since an element is vulnerable while moving on foot in enemy areas, it must use proper movement techniques and constant security to avoid unplanned enemy contact. This section discusses techniques applicable to the following areas dealing with foot movement:

- General principles.
- Techniques of movement for small units.
- Departure from and reentry into friendly areas.
- Route selection.
- Control measures.
- Selection and use of rallying points.
- Action at danger areas.
- Action on enemy contact (see Appendix B).

GENERAL PRINCIPLES

Regardless of the means of transportation into hostile territory, the following principles apply.

Know How To Navigate

Preparation and plans are worthless if the element cannot find its objective or, worse yet, stumbles into danger because of poor navigation. Plan to use at least two compass men and two pace men per element. Additionally, consider all aids to navigation, for example radar, RSTA, marking rounds, and guidance from the air.

Avoid Detection

Move by stealth and exploit the cover and concealment of the terrain. Move when visibility is reduced, such as during darkness, fog, snow, or rain. Use rough, swampy, or heavily vegetated terrain to help hide from the enemy. Exploit known weaknesses in enemy detection capabilities and plan movements when other operations divert the enemy's attention.

4-1

Maintain Constant Security

Even with well thought-out plans for movement, the element must take both *active* and *passive* security measures at all times. Each element member or subunit is responsible for security en route, at danger areas, at clandestine patrol bases, and most importantly, in the objective area.

Plan for Use of Supporting Fire

Plan for fire support (artillery, tactical air, attack helicopters, naval gunfire) even if it may not be needed during movement. A fire plan provides a tool to help move or navigate. For example, planned fire at known points along the route aids navigation. Detection can be avoided by planned fire that can destroy known enemy sensor fields or observation posts. Also, planned fire can be used to divert the enemy's attention away from an area through which the element is moving. Planned fire is ready to engage any threat, and it can be used to cover withdrawal from the objective area.

SMALL UNIT FOOT MOVEMENT

Special Forces and indigenous elements may move or maneuver as patrols using conventional techniques. These techniques are fundamental and can be adapted to element configuration. The patrol will be used as the vehicle throughout this chapter to illustrate these principles. Fire teams travel in wedges, and the larger unit moves on a column axis.

The enemy situation determines which of the three movement techniques to use. When contact is not likely, the patrol uses *traveling*. When contact is possible, *traveling overwatch* is used. When contact is expected, *bounding overwatch* is used. Patrols usually move by traveling overwatch because they are usually behind enemy lines and contact is possible.

In open terrain, the patrol leader should keep men widely dispersed. When enemy contact is possible, he should have one fire team in overwatch well forward of the other fire team. He should assign duties for the movement as shown in Figures 4-1, 4-2, 4-3, and 4-4.

Fire teams maintain visual contact, but the distance between them is such that the entire team does not become engaged if contact is made. Fire teams can spread their formations as necessary to gain better observation to the flanks (Figure 4-5). Although widely spaced, the men retain their relative positions in their wedge and follow their patrol

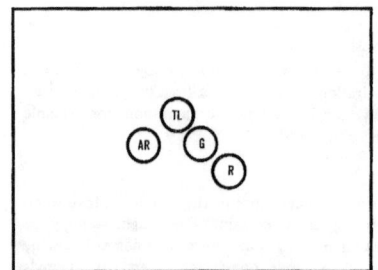

Figure 4-1. Fire team wedge.

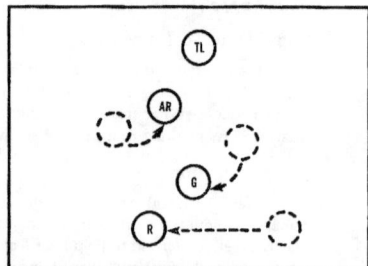

Figure 4-2. Modified fire team wedge.

leader. Terrain or weather may make it necessary to modify the wedge. Extreme situations may require movement in a single file(Figure 4-6).

The lead squad must secure the front. It should be the one best qualified to navigate and provide forward security for the patrol while en route. For a long movement, the patrol leader may rotate the lead squad. The fire team or squad in the rear is charged with rear security.

Movement of a platoon-size patrol is the same as in any other operation except that in platoon traveling overwatch the lead squad always moves by *squad* bounding overwatch to give that squad more security(Figure 4-7).

The patrol leader should vary movement techniques to meet the changing situation. If he needs to put the lead squad into traveling overwatch and have the patrol (-) overwatch the lead squad he should do it. This may be good for crossing a large open area.

Leaders, except fire team leaders, move within the formation where they can best control the situation and do their jobs. They can shift their men around. For instance, a patrol leader may want to have the pace men walk near him so that he can get an accurate distance report quickly(Figure 4-8).

In movement to contact, the patrol leader keeps key weapons with him for quick employment, as contact is most likely to come from the front. However, when patrolling, the patrol leader may place the weapons differently. For instance, he may have one machine-gun team forward under his control and one near the rear of the column with the assistant patrol leader(Figure 4-9), or he may have all key weapons move as a separate squad.

On some missions the patrol leader may want the assistant patrol leader to move with rather than in front of the trail squad to aid in control.

DEPARTURE FROM AND REENTRY INTO FRIENDLY AREAS, LINES, OR PERIMETERS

Other than when a patrol departs from within its own unit's portion of the forward edge of the battle area (FEBA), movement in forward unit areas must be controlled, coordinated, and kept to a minimum to avoid conflict with friendly troops or the activation of their RSTA devices. Forward units' positions are danger areas that must be assumed to be under enemy surveillance during all weather and visibility conditions. Patrols departing from their own unit's positions (Figure 4-10) have the advantage of more positive control and a less likely chance of being fired on by friendly troops.

Principles

When a patrol departs or reenters friendly areas, the following three principles apply.

Coordination. The patrol leader or his representative must coordinate departure and reentry of friendly areas either directly with the unit(s) through which the patrol will pass, or through designated staff agencies.

Reconnaissance. The patrol leader should reconnoiter (ground) the area through which the patrol will pass and return, and pick an IRP. He should observe the area just forward of the friendly unit for possible routes, danger areas, and obstacles. For this reconnaissance or coordination, the patrol should wear the same uniform as the men in the forward unit to avoid attracting the attention of enemy observers.

4-3

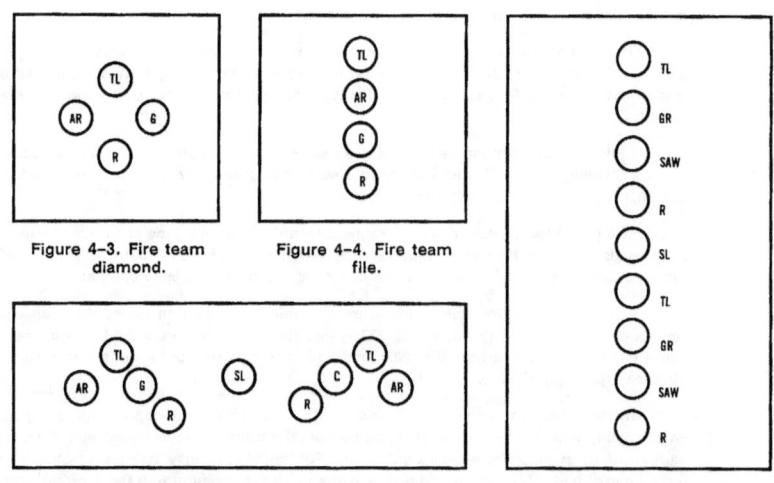

Figure 4-3. Fire team diamond.

Figure 4-4. Fire team file.

Figure 4-5. Squad line, wedge.

Figure 4-6. Squad file.

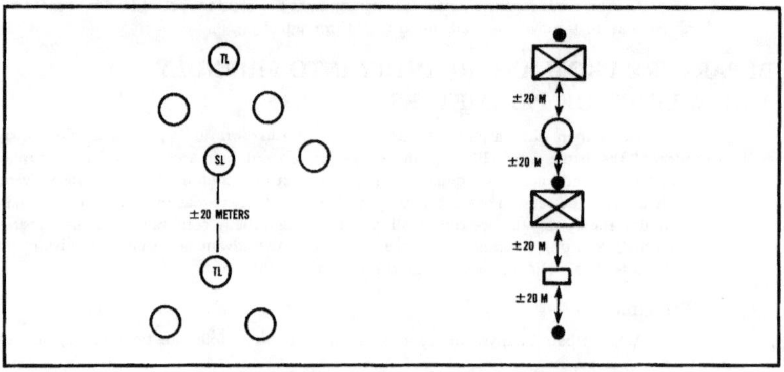

Figure 4-7. Traveling and traveling overwatch.

Security. The patrol should maintain security to avoid contact with the enemy while departing or reentering through friendly units. It is extremely important that the patrol make no enemy contact at this time as it is then so vulnerable. Control is very difficult if a fire fight starts just forward of a friendly forward unit.

4-4

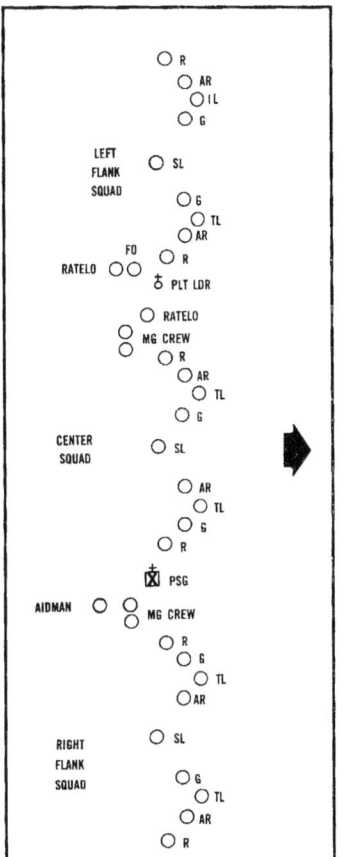

Figure 4-8. Platoon line, squads on line.

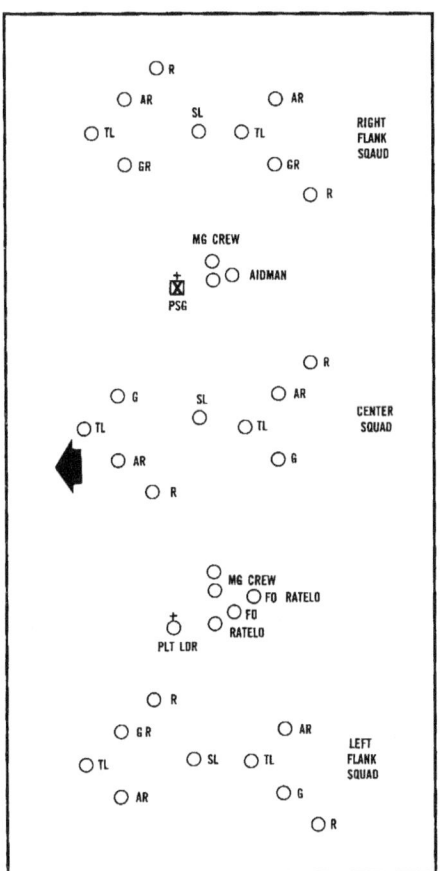

Figure 4-9. Platoon line, squads in column.

Techniques for Departure From Friendly Areas

The patrol leader establishes an IRP. (The selection and use of rallying points are explained on page 4-14.) The IRP may be occupied or just planned for, but all patrol members must know its location.

Security is maintained. Either a clearing party or an appropriate movement formation is used for departing the friendly unit.

4-5

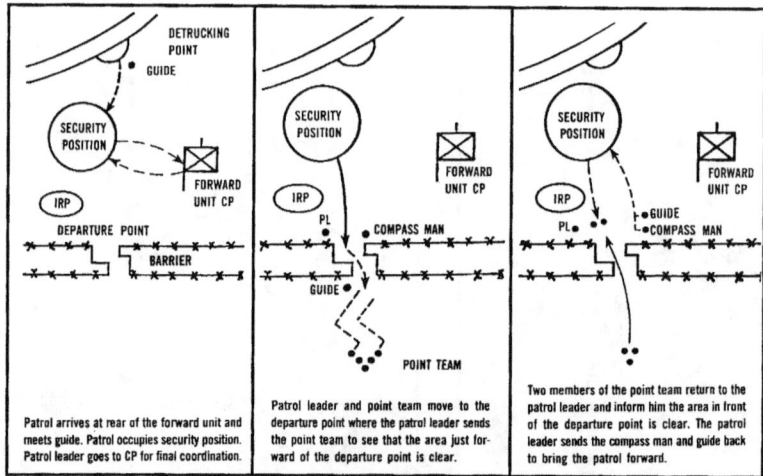

Figure 4–10. Departure from friendly lines.

Members of the patrol must not move within the friendly unit's area without a guide to lead them.

The patrol leader coordinates with the friendly unit commander to ensure no changes have occurred since initial coordination was made (this may be the first coordination made with the friendly unit).

Each patrol member is counted as he leaves (usually by the assistant patrol leader) to ensure all patrol members depart the friendly unit.

The patrol makes a security or listening halt for all members to adjust to the sights, sounds, and smell of the battle area. This halt is normally made beyond the friendly unit's final protective fires.

The patrol arrives at the forward unit and is met by a guide provided by the forward unit. The guide leads the patrol to the IRP chosen by the patrol leader during initial coordination with the forward unit commander.

No one should move, either singly or together, anywhere in the forward unit's area without a guide.

The patrol leader should then make final coordination with the forward unit commander. At this time, he learns of any changes that have taken place since first coordination and of any recent enemy activity that might affect the patrol.

Prior to leaving the patrol, the patrol leader gives instructions (called a contingency plan) for what should be done while he is gone. These instructions state where he is going, who he is taking with him, how long he will be gone, what to do if he does not return, and actions to take if there is enemy contact. If all goes well, he need not reissue these instructions when he leaves the patrol for final coordination.

4-6

80

On returning from final coordination, the patrol leader may issue a fragmentary order to cover any changes.

The technique for departing friendly areas depends on the enemy situation. There are common threats and techniques for countering the enemy: *ambush and chance contact, indirect fire*, and *RSTA*.

Ambush and chance contact. If the patrol leader learns that the enemy has patrols forward of friendly units, he must take steps to avoid enemy contact while departing.

- The patrol leader goes to the friendly side of the departure point where he dispatches the point team to see that the area forward of the barriers is clear.

- The point team checks the area out to the first covered position or to an area large enough to allow the patrol room to maneuver while departing. Distance varies according to the size of the patrol and the terrain.

- The point team notifies the patrol leader when the area is clear, and the rest of the patrol is brought forward.

Indirect fire. If enemy indirect fire is falling, the patrol leader should not halt his patrol after final coordination at the forward command post (CP), but he should be ahead of the main body far enough to provide security from ambush or chance contact.

Reconnaissance, surveillance, and target acquisition. If the enemy has RSTA devices, such as radar, sensor, or night vision devices, the patrol leader should counter this threat by taking the following countermeasures as appropriate:

- Use a well-hidden departure point such as a reverse slope or dense woods.

- Infiltrate the patrol through the departure area and have the men gather at a secure rendezvous point.

- Pass through the departure area when rain, fog, or snow helps to conceal the passage.

- Employ electronic countermeasures.

The patrol will have a security or listening halt after it has moved out of sight and sound of the forward unit. This is a short halt to see if the enemy is in the area, and to accustom the patrol to the sights and sounds of the battlefield. The patrol halts in a position that covers it from chance friendly small-arms fire. If the security or listening halt must be in an exposed position, each man gets down on one knee.

During a patrol, the patrol leader conducts frequent security or listening halts to ensure that the patrol is not followed and that no enemy is in the area.

Techniques for Reentry Into Friendly Areas

During reentry into friendly areas (Figure 4-11), the patrol leader—

- Establishes and occupies a reentry rallying point (RRP). (See page 4-14.)

- Maintains security at the RRP and at the reentry point.

- Uses as few men as possible to locate the reentry point while the bulk of the patrol remains in the RRP.

- Meets a guide at the reentry point. Normally, a password precoordinated forward of front lines is used since it may be overheard by the enemy.

- Ensures someone (usually the assistant patrol leader) accounts for the returning patrol members to ensure no infiltrators follow the patrol, especially during reduced visibility.

- Gives friendly unit commander a spot report providing only combat information of immediate tactical value to him.

- Moves the patrol into a rallying point near the reentry point. This rallying point should be on a prominent terrain feature where the patrol leader can pinpoint his location, with respect to the reentry point.

- By radio, alerts the forward unit that the patrol is ready to reenter. He uses a code word for security and brevity. The code word must be acknowledged by the forward unit before the patrol reconnoiters for the reentry point. This indicates that a guide has been sent to the reentry point and is waiting for the patrol.

- If the patrol leader is certain of the reentry point location, he moves the entire patrol directly to the reentry point and has the point team coordinate the reentry.

An effective method of locating the reentry point is through the use of a thermal imagery device, such as the AN/PAS-7. It can detect the body heat of the reentry guide, even though he may be well camouflaged, or his location masked by smoke. Also, if ground surveillance radar is available, it can be used to vector the patrol to the reentry point.

At no time should the reconnaissance patrol move parallel to friendly barriers or probe around the wire. If the reentry point cannot be found by the initial reconnaissance, or if the reconnaissance patrol comes into contact with the barrier wire, the patrol should notify

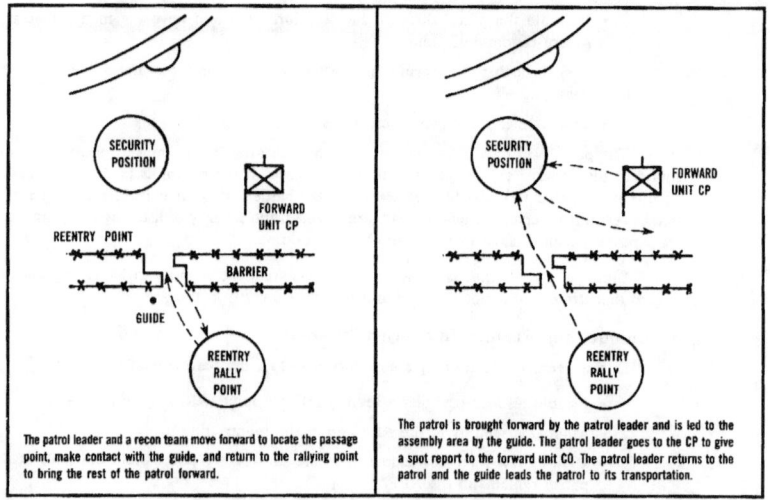

The patrol leader and a recon team move forward to locate the passage point, make contact with the guide, and return to the rallying point to bring the rest of the patrol forward.

The patrol is brought forward by the patrol leader and is led to the assembly area by the guide. The patrol leader goes to the CP to give a spot report to the forward unit CO. The patrol leader returns to the patrol and the guide leads the patrol to its transportation.

Figure 4-11. Reentry into friendly lines.

4-8

higher headquarters and move to another rallying point to wait until daylight. The patrol should *not* stay in a rallying point from which a radio transmission has been made.

When the reentry point is found, the patrol leader can either go back and bring the patrol forward, or he can call the assistant patrol leader and have him bring the patrol forward, provided the reentry point is easy to find.

The guide leads the patrol through the barriers to the security position previously coordinated with the forward unit commander. The patrol halts in the security position, remaining in movement formation. The patrol leader then gives a spot report to the forward unit commander to tell him anything of intelligence value or of immediate tactical use to the forward unit.

ROUTE SELECTION

The patrol picks routes to avoid contact with the enemy, local inhabitants, built-up areas, and natural lines of drift. All patrols, except those with a mission of attacking targets of opportunity, strive to reach their objectives without being detected. Selecting primary and alternate routes and dividing each route into legs (Figure 4-12) will help the patrol to remain undetected.

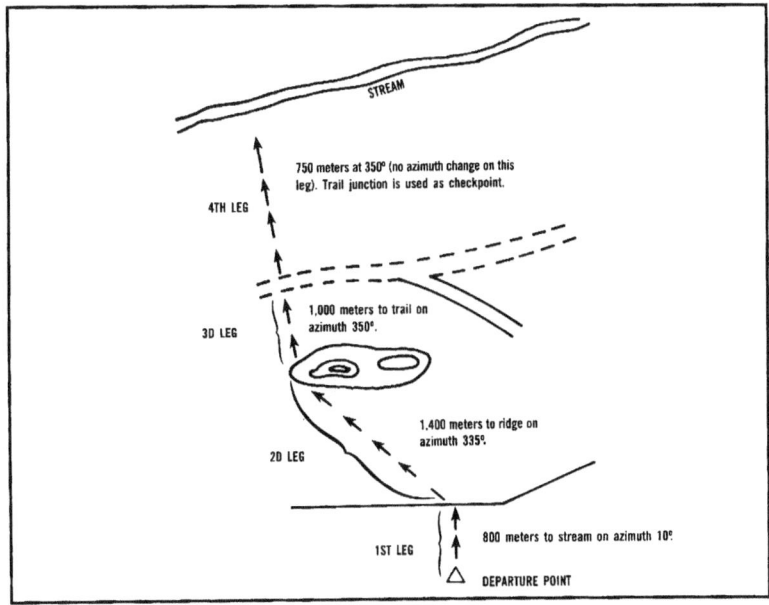

Figure 4-12. Divide route into legs.

4-9

Terrain Analysis

To pick a route, the patrol leader should analyze the terrain in the area in which the patrol will operate. One way to do this is to actually walk the ground. Another way is to make an aerial reconnaissance. If the tactical situation and the lack of aircraft preclude both ways, the patrol leader must make his analysis from a map and aerial photograph study. While maps and photos cannot give all the information sought, they can provide essential information.

The patrol leader analyzes the terrain in terms of its tactical aspects: observation and fields of fire, cover and concealment, obstacles, key terrain, and avenues of approach (OCOKA). He looks for danger areas and terrain features that will help navigation.

Observation and fields of fire. He should seek routes that afford the patrol good observation and (where necessary) good fields of fire. He should avoid areas that may expose the patrol to the enemy.

Cover and concealment. Cover and concealment are especially important to help avoid detection by the enemy.

Obstacles. He should seek routes that are free of obstacles that may impede the patrol's movement. He should also note obstacles that could be used to block an enemy attack or pursuit.

Key terrain. He should look for key terrain as the enemy may have it occupied or covered by fire.

Avenues of approach. He should avoid likely avenues of approach. The enemy may have them under visual surveillance or covered by fire or RSTA devices.

Tactical Considerations

In selecting the route, the patrol leader must consider the nature of the mission, time limitations, and the size and type (mounted or dismounted) of the patrol. He must also consider the danger of enemy contact, which may compromise the mission. Therefore, he should avoid known and suspected enemy locations by—

- Not picking a route parallel to enemy positions, as this increases the chance of discovery.

- Avoiding roads and trails, as they are danger areas (likely places for ambush or chance enemy contact).

- Avoiding all built-up areas. Regardless of the sympathies of the local population, one loose word or one enemy sympathizer could mean disaster for a patrol. Domestic animals may also alert the enemy to the patrol's presence.

During daylight, the patrol should use routes concealed by heavy vegetation to protect it from enemy observation. During darkness, it should use a route that affords silent movement.

Natural obstacles such as swamps and cliffs can hinder the patrol's movement, but they can also be used to gain surprise if the enemy concentrates on more likely avenues of approach. The patrol should avoid man-made obstacles (barbed wire, craters, roadblocks, minefields), as they may be covered by enemy fire or detection devices.

The patrol leader should plan the route to avoid being detected by known or suspected enemy RSTA devices.

Navigational Considerations

The patrol leader should select *prominent terrain features* along the route and memorize their locations. Use these features as checkpoints and to help divide the route into legs. Legs should be manageable, that is, not too short nor too long. Terrain is a major consideration for length. A leg requires only a terrain feature, not necessarily an azimuth change.

Two helpful navigational techniques are the offset–compass method and the box–in method. These methods should be combined.

Offset–compass method. This method is a planned deviation to the right or left of a straight line azimuth to the patrol's destination. Each degree of offset moves the patrol 17 meters right or left for each kilometer traveled.

Box–in method. This method uses natural or man–made features, such as roads or streams that form boundaries for a route. By referring to these boundaries, the patrol can recognize and correct any large deviation from the planned route.

Route Selection in Different Types of Terrain

Depending on the type of terrain, the patrol leader must consider cover and concealment possibilities when selecting a route.

Mountains. In selecting a route over mountainous terrain, the patrol leader must weigh the added security of steep ridges and cliffs against the disadvantage of tiring the patrol. Natural lines of drift, such as ridges, draws, and streams are characteristic of mountainous terrain and are difficult to avoid.

Desert. In the desert, patrol routes must exploit all concealment offered by any vegetation, shadows, or ravines (defilade). The patrol leader should plan routes for use at night which will bring the patrol to a concealed position before daylight. Water replenishment is a major factor in route planning.

Jungle. In a jungle, the topsoil is loose and soft. Slopes are slippery, and walking on them leaves an obvious trail. Routes should follow ridge lines or low ground where movement is faster, less tiring, and less likely to leave a trail. Because the few roads, trails, and rivers in a jungle are heavily used by civilians and the enemy, the patrol should avoid them when possible or cross them only when visibility is reduced.

Swamp. Normally, a patrol must—

- Use dead reckoning in navigating a featureless swamp.
- Plan the route to take advantage of swamp islands that can be used for clandestine patrol bases.
- Cross rivers and streams at a point below where branch streams join to avoid numerous crossings of the same stream.
- Cross rivers and streams under cover of darkness.

Arctic. As a rule in arctic terrain, the patrol follows features that are easiest to walk. In selecting a route in snow–covered terrain, the patrol leader should consider the followingtypes of terrain and water routes.

In *open terrain*, when feasible, break the trail along a tree line so that the shadows can help to conceal the trail and the patrol moving on it. Select rough ground to use available shadows to conceal tracks and patrol members. (When you have a tree line you have woods and if you have woods, you have concealment and don't have *open terrain!*)

In *covered terrain*, when feasible, the trail should be in a forest (with little or no underbrush) that provides concealment and protection from the wind. Thickets and windfall forest areas require work to break a trail through them. Avoid them.

In *hilly and mountainous terrain*, valleys and frozen rivers most often provide the easiest route in snow–covered country. If a valley cannot be used, the trail may be broken down on the lee side (away from the wind) of a ridge line or hill mass that dominates the valley. Avoid areas of avalanche conditions. Use gentle inclines when climbing or descending.

Tree–lined frozen lakes, rivers, and creeks ease navigation and offer suitable *water routes* in arctic areas. For protection and concealment, ski close to the bank to permit quick movement into wooded areas on shore. Check thickness of ice before using any ice route. The minimum thickness for one rifleman on skis is 5 centimeters (2 inches); for an infantry column in single file on foot, 10 centimeters (4 inches). Warm water springs, prevalent in northern areas, may create a hazard to both foot and vehicle movement. Many of these springs do not freeze, even in extremely low temperatures. They may cause streams to have little or no ice and some lakes to have only thin ice. Their presence in muskeg or tundra can cause weak spots in otherwise trafficable terrain. These areas should be either bridged, reinforced, or bypassed.

Selection of Alternate Routes

As a rule, the patrol leader selects one route to the objective, a different route to return to friendly areas (to reduce the chance of ambush), and one alternate route that may be used either to or from the objective. This allows flexibility to meet the changing tactical situation.

The patrol should use the alternate route when it has had contact with the enemy on the primary route, or when for some other reason the patrol has been detected. An alternate route must be—

- One with the same tactical and navigational characteristics as the primary route.
- Far enough away from the primary route so that movement on both routes cannot be detected from one position.
- Coordinated in the same way and at the same time as the primary route.

CONTROL

The success of a patrol depends in large part on how well the patrol leader controls it. He must control its direction and speed of movement, and he must be able to start, stop, or shift fire if needed. All patrol members can help control by staying alert and passing signals and orders on to others. The subordinate leaders move with and control their elements. They remain alert for signals and orders and ensure their men receive and comply with them.

Signals

The patrol leader can use audio, visual, and physical signals to maintain control.

Audio. Audio signals such as voice, radios, whistles, or field phones can be used. However, audio signals may have limited use depending on the situation.

Voice is a good means of control but must frequently be kept to a whisper.

Radios are good for control, especially in a large patrol, if good radio discipline is practiced. However, the patrol leader must always assume the enemy has the capability to intercept and locate the patrol with direction finding equipment. It can then maneuver its troops to attack the patrol, fire artillery, or at least alert its rear–area troops to the presence of the patrol behind its lines. Radio can, therefore, be the least desirable means. Radio transmission must be short and if possible, use one–directional antennas (which are easy to make with field expedients). During a raid, however, radios are an excellent means of initiating the support element's fire, directing artillery, and controlling the raid. Once a fight begins, the danger of radio intercept becomes unimportant.

Whistling is another means to signal such actions as a withdrawal from the objective. A whistle is not a good signaling device when secrecy is necessary. It may be difficult to hear over the sound of gunfire, and it may be ineffective in halting or shifting fire.

Field phones are useful in a stationary position, such as a patrol base or ambush site. The patrol leader should consider the weight of such equipment when planning.

Visual. Visual signals may be used to halt a patrol, to indicate detection and direction of the enemy, or to indicate that all is clear. However, like audio signals, visual signals may also have limited use.

Pyrotechnics are good for signaling as they attract attention and can be seen a long way. However, they can also be seen by the enemy and used to pinpoint the position of the raid. Varying combinations should be used and patterns should be avoided.

Arm–and–hand signals may be used whenever possible instead of the radio or voice, especially when close to the enemy. In addition to standard signals (FM 21–60), the patrol leader can devise any arm–and–hand signals he wants to use within the patrol. They must be simple, easily understood, appropriate, and well rehearsed.

Luminous tape on a patrolling cap or the *luminous marks* on a compass may be used at night, over short distances, as signals.

Infrared sending and receiving equipment, such as the sniperscope, the infrared weapon sight, the metascope, night vision goggles, and the infrared filters for the flashlight can be used to send and receive signals at night. Other items of RSTA equipment can be used in much the same way. However, RSTA equipment is bulky and may be difficult to get. The patrol should take the same precautions with it as with radios and pyrotechnics, as the enemy may have similar equipment.

Physical. A *tugline* is a reliable and secure method of signaling. By tying a string, rope, or wire from one man to another, signals can be passed quietly and quickly by pulling on the wire in a prearranged code. Tuglines are difficult to install but they can be used in a static position, such as a patrol base or ambush site.

Time

Time is a good secondary means of control. By giving a person a time schedule for certain actions, these actions can be controlled.

Accounting for Patrol Members

When a patrol moves over long distances and through difficult terrain patrol members after crossing danger areas, after enemy contact, after crossing an obstacle, after halts, and periodically while moving.

When moving in a modified wedge, the last man sends up the count by tapping the man in front of him and saying "one" in a whisper. This man taps the man in front of him and says "two". This continues until the count reaches the patrol leader. Each man makes sure that the man he taps receives and passes on the count.

In a large patrol, the patrol leader may need to halt and have the subordinate leaders check their men and make a report.

Instructions to subordinate leaders or to the last man in the patrol as to when to send up the count will keep the patrol leader from having to ask for it.

Control of the Point Team

The mission of the point team is to provide essential frontal security. The team leader can be responsible for navigation of the patrol. Another technique is that the team maintains a general azimuth and guides on the patrol as necessary.

The point team should be positioned far enough to the front to provide early warning of danger areas and to allow the main body sufficient room to maneuver if the point encounters the enemy. This may vary from a few meters to several hundred.

The primary means of control is to maintain visual contact between the point team and main body at all times. Through arm-and-hand signals, the patrol leader can pass orders to the point team leader. In addition to visual means, the point team can be required to make physical contact with the main body after a specified time or distance. Radios should be used only in an emergency.

SELECTION AND USE OF RALLYING POINTS

A rallying point is a place where a patrol can—

- Reassemble and reorganize if dispersed.

- Temporarily halt to reorganize and prepare for action at the objective.

- Temporarily halt to prepare for reentry into friendly areas.

A rallying point should—

- Be easily recognizable.

- Have cover and concealment.

- Be defensible for a short period of time.

- Be away from natural lines of drift.

When planning a patrol, the patrol leader makes a thorough map reconnaissance to pick areas likely to be designated as rallying points. The ORP, the IRP, and the RRP are designated in the patrol order. Rallying points en route are normally not designated in the patrol order. However, if the mission is complex and the distance to the objective is long, designate tentative rallying points in the patrol order by terrain features and their grids.

A *rallying point* is a location the patrol physically passes through. A *rendezvous point* is a rallying point not physically passed through.

Tentative rallying points may be designated for extended operations if the patrol is dispersed and unable to assemble at a previously designated rallying point.

If enemy activity precludes the use of the last designated rallying point, the patrol uses the one previously designated.

A time limit for reassembly and actions to be taken in a rallying point must be specified in the patrol order.

The ORP will be used as a rallying point when tentative rallying points have not been designated in the operation order. If any patrol member reaches the rallying point designated on the ground after the time limit has elapsed, he must strive to rejoin the patrol at the ORP.

The following are two of many techniques for designating rallying points and for informing all patrol members. Variations of these techniques can be developed.

If visibility permits, the patrol leader designates rallying points by arm–and–hand signals. This technique is especially good for small patrols.

If the patrol is spread out so that all patrol members cannot use the patrol leader's arm–and–hand signals, the patrol leader should halt the patrol, have the point team leader and assistant patrol leader come to his location, and tell them that he is designating that place as a rallying point. The assistant patrol leader should then stay at the rallying point and tell each member of its designation as he passes. The point team leader returns to his team and tells it of the rallying point location.

Normally, en route and known danger area rallying points are planned for and designated while moving.

DANGER AREAS

There are four types of danger areas: linear, small open, large open, and series.

A *linear* danger area is best characterized by roads and trails. Each flank of the patrol is exposed to a relatively narrow field of fire. However, streams often afford the same advantages to an enemy as a road.

A *small open* danger area is an open area of such a size that the patrol can be hit in one flank and/or its front by enemy small–arms fire.

A *large open* danger area is an open area of such size that the lead team of the patrol is beyond effective small arms of the overwatch element.

A *series* of danger areas can be similar to a large open danger area, especially when the danger areas are a series of linear danger areas. However, these danger areas can also be enemy defensive strong points and trench lines.

The patrol should cross a danger area where observation is restricted, such as a curve in the road, or where vegetation comes right up to both sides of the road.

It should secure the near side of the danger area and secure flanks. Usually a visual reconnaissance and the presence of the patrol is enough to secure the near side.

The patrol leader should designate *near side* rallying points and *far side* rendezvous points, if they are not already designated in the operation order (OPORD). If they are designated, but the patrol leader decides to change one or more points on arrival at the danger area, he must inform all patrol members before crossing. The rallying point on the *near* side is usually the last rallying point designated before encountering the danger area. The rendezvous point on the *far* side must be a safe distance on the far side of the danger area along the route of march.

4-15

The patrol should reconnoiter and secure the far side. This may require that some members cross the danger area to see if the enemy is there and to see if the crossing site is safe and suitable. However the patrol leader may decide that a visual reconnaissance of the far side is adequate. In either case, the patrol should not cross the danger area until the reconnaissance is complete.

If a patrol is split by enemy action while crossing, patrol members who have already crossed should go to the rendezvous point on the far side. Those members who have not crossed should return to the rallying point on the near side. At the rallying point on the near side, the senior patrol member must take command and try to cross the danger area at another place. If, after crossing the danger area, the time limit has not expired, he must try to rejoin the remainder of the patrol at the rendezvous point on the far side. If the time limit has expired, he must try to rejoin the patrol in the ORP or tentative rallying point, based on the instructions issued in the patrol order.

The patrol must remove evidence (such as footprints) of its crossing the danger area.

When crossing linear danger areas (for example, roads, trails, and streams), the point team alerts the patrol leader that a danger area is to the front.

The patrol leader goes forward to see if he should proceed with his original plan or modify it. He also decides if the tentative rallying point on the near side and the rendezvous point on the far side are suitable.

The point team reconnoiters and secures the near side (Figure 4-13). Security teams are then sent to the flanks. These teams should be able to communicate with (signal) the patrol.

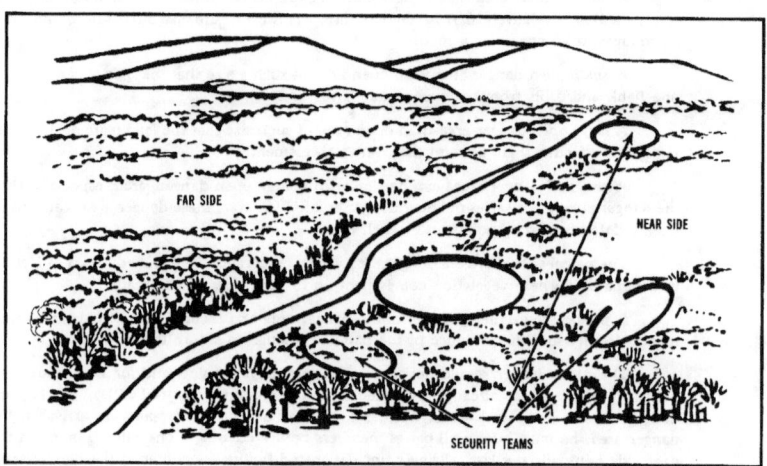

Figure 4-13. Securing near side.

4-16

The patrol leader also considers terrain and visibility when crossing linear danger areas. He puts security teams out far enough to give the patrol warning of enemy approach and to prevent the patrol from being hit by enemy fire directed at a security team.

When flank security is in place, he sends the point team across the danger area to reconnoiter and secure the far side (Figure 4-14). A small patrol, such as a five-man patrol may not be able to put a team on each flank.

The point team reconnoiters forward, far enough to allow room for the patrol to move back into its formation.

There are several methods for the main body of a patrol to cross. It may use a culvert, it may cross in groups, or it may use the modified wedge (file) or on line.

The point team stays on the far side of the danger area where it is least exposed. This is a reason for picking a place where there is thick vegetation right up to the edge of the danger area on both sides.

The assistant patrol leader ensures that the flank security rejoins the patrol. Each flank security team moves directly across the danger area (Figure 4-15) and then moves diagonally to rejoin the patrol. The patrol leader should slow the rate of movement, or even halt the patrol, so the flank security teams can rejoin the patrol.

The patrol should bypass small open areas whenever possible (Figure 4-16). As with any danger area, when the patrol leader is alerted by the point team that an open area is to the front, he will go forward to see if he should proceed on the original route or bypass it.

The patrol uses the techniques for small open areas and linear danger areas when crossing large open danger areas (Figure 4-17). The patrol leader has the option of changing the original plan. Consider the feasibility of bypassing. Figure 4-17 shows the patrol crossing a large open area when the area cannot be bypassed.

To reach the enemy's rear area, the patrol may have to pass through the enemy's MLD. The enemy's MLD is similar to our FEBA in that it is a line of its forwardmost defense positions. It is defended in depth, and may include foxholes, bunkers, wires, minefields, and other obstacles.

A successful passage requires terrain or enemy disposition, or both, that permits the patrol to move through the enemy's forward positions without detection. If the enemy is alert, it may prevent passage. However, the patrol leader should always consider the technique of passing through the enemy's MLD as a means of moving the patrol into the enemy's rear area.

Penetration of the enemy's MLD requires time, detailed planning, and thorough rehearsals. As movement is by stealth, it is slow.

The patrol leader should select routes for penetration that have considerable cover and concealment. Areas such as woods, swamps, and rough terrain, which limit enemy observation and use of surveillance devices, are best suited for passage. The patrol leader should seek a passage area where the enemy is dispersed with wide gaps between positions. Coordination with local ground surveillance radar (GSR) or REMS teams through the unit's S2 can assist in obtaining this information.

The patrol must not use main avenues of approach as the enemy will have them covered by observation or fire. If the enemy is alert and has detection devices, the patrol

4-17

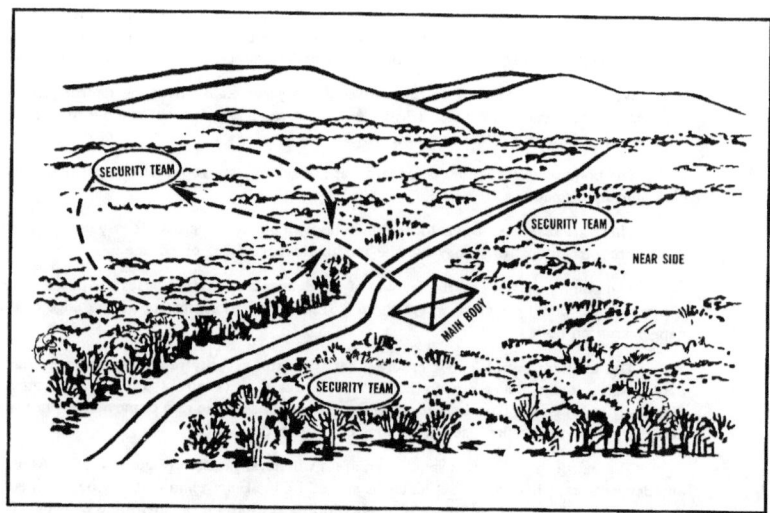

Figure 4-14. Securing far side.

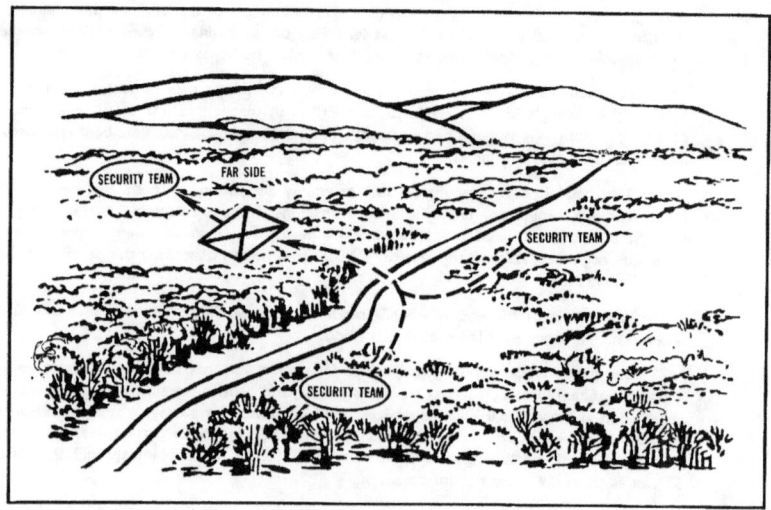

Figure 4-15. Crossing danger area.

4-18

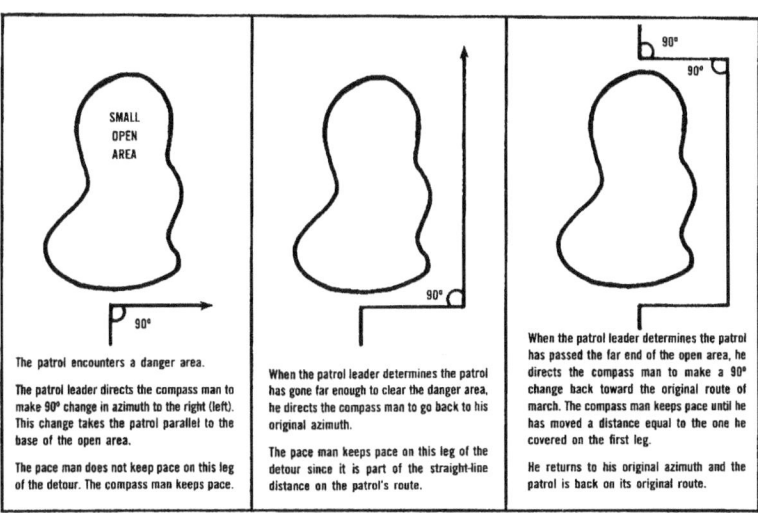

Figure 4-16. Small open areas.

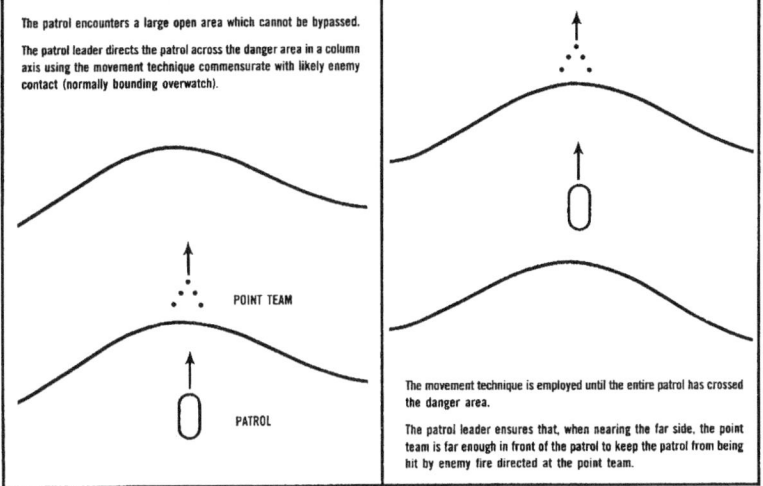

Figure 4-17. Large open areas.

4-19

must use deception and diversionary measures. Planned supporting fire covers and aids movement of the patrol.

The patrol must find a weak point in the enemy's MLD where the enemy's defenses are not fully developed or where a gap exists. In cases where no weak point exists, they may be created through combat action or deception.

A reconnaissance party, consisting of the patrol leader, compass man, and a two-man security team, moves by bounds probing to find a weak point. The patrol leader must tell the men in the reconnaissance party what to do if contact is made.

The reconnaissance party observes a sector of the MLD until it detects what it feels is a weak point.

The patrol leader returns to the patrol with the compass man, issues a fragmentary order (if necessary), and moves the patrol to the weak point. The formation that the patrol uses should present as small a frontal target as possible, but retain the ability to fire to the front and flanks.

The patrol moves through the weak point quickly and stealthily.

If the patrol is detected, it can pull back or push on through depending on which gives it the best chance to survive and succeed. It may call for fire support.

ACTIONS ON ENEMY CONTACT

Unless it is required by its mission, a patrol must strive to avoid enemy contact. However, a patrol may unintentionally contact enemy troops. It must then quickly break contact so it can continue on its mission. If the situation allows, the patrol leader will quickly plan and give orders (fragmentary) to break contact and extract the patrol.

In some unintentional contacts, such as an ambush, there is not time for planning and giving orders. The patrol must take immediate fast action according to a well-rehearsed plan called an *immediate action drill*. There may be several of these drills planned with one for each type situation likely to be encountered (see Appendix B). They should not be put to such frequent and routine use that the enemy can predict and counter them.

Situations calling for immediate action drills also call for aggressive, sometimes violent execution.

As soon as the patrol leader or any member of the patrol recognizes a situation requiring it, he initiates (signals for) the appropriate immediate action drill. Instantaneous action usually gives the best chance for success and survival.

Immediate action drills can be developed by any unit or patrol, no matter how it is organized. They should be developed to fit the type terrain, the enemy's practices, and the mission of the patrol or type operations of the friendly unit. They are used to—

- Counter an ambush.
- React to unintentional contact at close range when terrain restricts maneuver.
- Defend against low-level air attack.

Section II. Stay–Behind Operations

Stay–behind operations involve the positioning of SF operational elements within their proposed operational areas prior to enemy advances through or occupation of general areas. Stay–behind operations should be considered when the civil populace indicates that it will support stay–behind operations.

ADVANTAGES AND DISADVANTAGES OF STAY–BEHIND OPERATIONS

Stay–behind operations enable SF elements to organize a nucleus of resistance forces. They also enable SF elements to pre–position themselves for employment in a unilateral role. Figure 4–18 illustrates other advantages, as well as some disadvantages of stay–behind operations.

ADVANTAGES	DISADVANTAGES
• Operations may be planned and rehearsed prior to hostilities.	• Enemy troops are nearby during occupation of the operational area.
• Less external support is required.	
• Personnel are familiar with the operational area.	• Movement and communications are greatly restricted.
• A high degree of security is possible.	
• Previously established civilian contacts may be exploited.	• Special Forces members may be compromised by informers.
• Caches of supplies and equipment can be established.	• Care and evacuation of the wounded hampers the mission and lend themselves to compromise.
• Immediate intelligence is available.	

Figure 4-18. Advantages and disadvantages of stay-behind operations.

OTHER CONSIDERATIONS

Special Forces elements must take stringent precautions to preserve security, particularly at refuge areas or at other safe sites to be used during the initial period of occupation. They must keep information concerning locations and identities within the organization on a need–to–know basis. Contacts between the various elements involved in a stay–behind operation are held to a minimum.

Commanders should consider pre–positioning personnel. This consists of placing highly trained and selected personnel in areas to function as intelligence agents and to establish and maintain contact with underground elements. When stay–behind operations are attempted, the SF elements may be completely dependent on the indigenous organization for security, for contacts required for expansion, and for buildup.

4–21

Pre-positioning supplies, especially medical supplies, may be required, notably in the likely event that the host nation does not have adequate technology to supply modern medical necessities.

SUBSEQUENT ACTIVITIES

Stay-behind elements may be used for intelligence missions and for demolitions that the withdrawing force was unable to execute. Subsequent activities begin at an appropriate time when civilian or military leaders call upon the population of the occupied area to continue resistance against the enemy. These subsequent activities include all forms of guerrilla warfare, subversion, and sabotage.

CHAPTER 5

CACHING

Caching is the process of hiding equipment or materials in a secure storage place with the view to future recovery for operational use. The ultimate success of caching may well depend upon attention to detail, that is, professional competence, that may seem of minor importance to the untrained eye. Security factors, such as cover for the caching party, sterility of the items cached, and removal of even the slightest trace of the caching operation are vital. Highly important, too, are the technical factors that govern the preservation of the items in usable condition and the recording of data essential for recovery. Successful caching entails careful adherence to the basic principles of clandestine operations, as well as familiarity with the technicalities of caching.

Section I. Caching Considerations

Caching considerations that are vital to the success of the caching operation may be done in a variety of operational situations. For example, cached supplies can meet the emergency needs of personnel who may be barred from their normal supply sources by sudden developments or who may need travel documents and extra funds for quick escape. Caching can help solve the supply problems of long-term operations conducted far from a secure base.

Caching also can provide for anticipated needs of wartime operations in areas likely to be overrun by the enemy.

PLANNING FOR A CACHING OPERATION

Caching involves selecting the items to be cached, procuring those items, and selecting a cache site. Selection of the items to be cached requires a close estimate of what will be needed by particular units for particular operations. Procurement of the items usually presents no special problems. In fact, the relative ease of procurement before an emergency arises is one of the prime considerations in favor of caching. When selecting a cache site, planners should always ensure that the site is accessible not only for emplacement, but also for recovery. When planning a caching operation, the planner must consider seven basic factors.

Purpose and Contents of the Cache

Planners must determine the purpose and contents of each cache because these basic factors influence the location of the cache, as well as the method of hiding. For instance, small barter items can be cached at any accessible and secure site because they can be concealed easily on the person once recovered. However, it would be difficult to conceal rifles for a guerrilla band once recovered. Therefore, this site must be in an isolated area where the band can establish at least temporary control. Certain items, such as medical stock, have limited shelf life and require rotation periodically or special storage considerations, necessitating easy access to service these items. Sometimes it is impossible to locate a cache in the most convenient place for an intended user. Planners must compromise between logistical objectives and actual possibilities when selecting a cache site. Security is always the overriding consideration.

5-1

97

Anticipated Enemy Action

In planning the caching operation, planners must consider the capabilities of any intelligence or security services not participating in the operation. They should also consider the potential hazards the enemy and its witting or unwitting accomplices present. If caching is done for wartime operational purposes, its ultimate success will depend largely on whether the planners anticipate the various obstacles to recovery, which the enemy and its accomplices will create if the enemy occupies the area. What are the possibilities that the enemy will preempt an ideal site for one reason or another and deny access to it? A vacant field surrounded by brush may seem ideal for a particular cache because it is near several highways. But such a location may also invite the enemy to locate an ordnance depot where the cache is buried.

Activities of the Local Population

Probably more dangerous than deliberate enemy action are all of the chance circumstances that may result in the discovery of the cache. Normal activity, such as construction of a new building, may uncover the cache site or impede access to it. Bad luck cannot be anticipated, but it can probably be avoided by careful and imaginative observation of the prospective cache site and of the people who live near the site. If the cache is intended for wartime use, the planners must project how the residents will react to the pressures of war and conquest. For example, one of the more likely reactions is that many residents may resort to caching to avoid having their personal funds and valuables seized by the enemy. If caching becomes popular, any likely cache site will receive more than normal attention.

Intended Actions by Allied Forces

Using one cache site for several clandestine operations involves a risk of mutual compromise. Therefore, some planners should rule out otherwise suitable caching sites if they have been selected for other clandestine purposes, such as drops or safe houses. A site should not be located where it may be destroyed or rendered inaccessible by bombing or other allied military action, should the area be occupied by the enemy. For example, installations likely to be objects of special protective efforts by the occupying enemy are certain to be inaccessible to the ordinary citizen. Therefore, if the cache is intended for wartime use, the caching party should avoid areas, such as those near key bridges, railroad intersections, power plants, and munitions factories.

Packaging and Transportation Assets

Planners should assess the security needs and all of the potential obstacles and hazards that a prospective cache site can present. They should also consider whether the operational assets of the organization are sufficient to overcome those obstacles and hazards securely. Planners must consider the assets that could be used for packaging and transporting the package to the site. Best results are obtained when the packaging is done by experts at a packaging center. The first question, therefore, is to decide whether the package can be transported from the headquarters or the field packaging center to the cache site securely and soon enough to meet the operational schedules. If not, the packaging must be done locally, perhaps in a safe house located within a few miles of the cache site. If such an arrangement is necessary, the choice of cache sites may be restricted by limited safe house possibilities.

Personnel Assets

All who participate directly in emplacement will know where the cache is located. Therefore, only the fewest possible and the most reliable persons should be used. Planners must consider the distance from the persons' residence to the prospective cache site and what action cover is required for the trip. Sometimes transportation and cover difficulties require the cache site to be within a limited distance of the persons' residence. The above considerations also apply to the recovery personnel.

Caching Methods

Which cache method to use depends on the situation. It is therefore unsound to lay down any general rules, with one exception. Planners should always think in terms of suitability, for example, the method most suitable for each cache, considering its specific purpose; the actual situation in the particular locality; and the changes that may occur if the enemy gains control.

Concealment. Concealment requires the use of permanent man–made or natural features to hide or disguise the cache. It has several advantages. Both emplacement and recovery usually can be done with minimum time and labor, and cached items concealed inside a building or dry cave are protected from the elements. Thus, they require less elaborate packaging. Also, in some cases, a concealed cache can be readily inspected from time to time to ensure that it is still usable. However, there is always the chance of accidental discovery in addition to all the hazards of wartime that may result in discovery or destruction of a concealed cache or denial of access to the site. The concealment method, therefore, is most suitable in cases where an exceptionally secure site is available or where a need for quick access to the cache justifies a calculated sacrifice in security. Concealment may range from securing small gold coins under a tile in the floor to walling up artillery in caves.

Burial. Adequate burial sites can be found almost anywhere. Once in place, a properly buried cache is generally the best way of achieving lasting security. In contrast to concealment, however, burial in the ground is a laborious and time–consuming method of caching. The disadvantages of burial are that—

- Burial almost always requires a high–quality container or special wrapping to protect the cache from moisture, chemicals, and bacteria in the soil.

- Emplacement or recovery of a buried cache usually takes so long that the operation must be done after dark unless the site is exceptionally secluded.

- It is especially difficult to identify and locate a buried cache.

Submersion. Submersion sites that are suitable for secure concealment of a submerged cache are few and far between. Also, the container of a submerged cache must meet such high technical standards for waterproofing and resistance to external pressure that the use of field expedients is seldom workable. To ensure that a submerged cache remains dry and in place, planners must determine not only the depth of the water, but the type of bottom, the currents, and other facts that are relatively difficult for nonspecialists to obtain. Emplacement, likewise requires a high degree of skill. At least two persons are needed for both emplacement and recovery. Especially when a heavy package is involved, recovery is often more difficult than emplacement and requires additional equipment. In view of the difficulties—especially the difficulty of recovery—the submersion method is suitable only on rare occasions. The most noteworthy usage is the relatively rare maritime resupply operation where it is impossible to deliver supplies directly to a reception committee. Caching supplies offshore by submersion is often preferable to sending a landing party ashore to bury a cache.

SELECTION OF THE SITE

The most careful estimates of future operational conditions cannot ensure that a cache will be accessible when it is needed. However, criteria for a site selection can be met when three questions are answered.

Criteria for Site Selection

Can the site be located by simple instructions that are unmistakably clear to someone who has never visited the location? A site may be ideal in every respect, but if it has no distinct, permanent landmarks within a readily measurable distance it must be ruled out.

Are there at least two secure routes to and from the site? Both primary and alternate routes should provide natural concealment so that the emplacement party and the recovery party can visit the site without being seen by anyone normally in the vicinity. An alternate escape route offers hope of avoiding detection and capture in an emergency.

Can the cache be emplaced and recovered at the chosen site in all seasons? Snow and frozen ground create special problems. Snow on the ground is a hazard because it is impossible to erase a trail in the snow. Planners must consider whether seasonal changes in the foliage will leave the site and the routes dangerously exposed.

The Map Survey

Finding a cache site is often difficult. Usually, a thorough systematic survey of the general area designated for the cache is required. The survey is best done with as large-scale a map of the area as is available. By scrutinizing the map, the planners can determine whether a particular sector must be ruled out because of its nearness to factories, homes, busy thoroughfares, or probable military targets in wartime. A good military-type map will show the positive features in the topography: proximity to adequate roads or trails, natural concealment (for example, surrounding woods or groves), and adequate drainage. A map also will show the natural and man-made features in the landscape. It will provide the indispensable reference points for locating a cache site: confluences of streams, dams and waterfalls, road junctures and distance markers, villages, bridges, churches, and cemeteries.

The Personal Reconnaissance

A map survey normally should show the location of several promising sites within the general area designated for the cache. To select and pinpoint the best site, however, a well-qualified observer must examine each site firsthand. If possible, whoever examines the site should carry adequate maps, a compass, a drawing pad or board for making sketch maps or tracings, and a metallic measuring line. (A wire knotted at regular intervals is adequate for measuring. Twine or cloth measuring tapes should not be used because stretching or shrinking will make them inaccurate if they get wet.) The observer should also carry a probe rod for probing prospective burial sites, if the rod can be carried securely.

Since an observer seldom completes a field survey without being noticed by local residents, his action cover is of great importance. His cover must offer a natural explanation for his exploratory activity in the area. Ordinarily, this means that an observer who is not a known resident of the area can pose as a tourist or a newcomer with some reason for visiting the area. However, his action cover must be developed over an extended period before he undertakes the actual reconnaissance. If the observer is a known resident of the area, he cannot suddenly take up hunting, fishing, or wildlife photography without arousing interest and perhaps suspicion. But he must build up a reputation for being a devotee of his sport or hobby.

Reference Points

When the observer finds a suitable cache site, he prepares simple and unmistakable instructions for locating the reference points. These instructions must identify the *general area* (the names of generally recognizable places, from the country down to the nearest village) and an *immediate reference point*. Any durable landmark that is identified by its title or simple description can be the immediate reference point (for example, the only Roman Catholic Church in a certain village or the only bridge on a named road between two villages). The instructions must also include a *final reference point (FRP)*, which must meet four requirements:

- It must be identifiable, including at least one feature that can be used as a precise reference point.
- It must be an object that will remain fixed as long as the cache may be used.
- It must be near enough to the cache to pinpoint the exact location of the cache by precise linear measurements from the FRP to the cache.
- It should be related to the immediate reference point by a simple route description, which proceeds from the immediate reference point to the FRP.

Since the route description should be reduced to the minimum essential, the ideal solution for locating the cache is to combine the immediate reference point and the FRP into one landmark readily identifiable, but sufficiently secluded. The following objects, when available, are sometimes ideal reference points: small, unfrequented bridges and dams, boundary markers, kilometer markers and culverts along unfrequented roads, a geodetic survey marker, battle monuments, and wayside shrines. When such reference points are not available at an otherwise suitable cache site, natural or man-made objects may serve as FRPs: distinct rocks, posts for power or telephone lines, intersections in stone fences or hedgerows, and gravestones in isolated cemeteries.

Pinpointing Techniques

Recovery instructions must identify the exact location of the cache. These instructions must describe the point where the cache is placed in terms that relate it to the FRP. When the concealment method is used, the cache ordinarily is placed inside the FRP, so it is pinpointed by a precise description of the FRP. A submerged cache usually is pinpointed by describing exactly how the moorings are attached to the FRP. With a buried cache, any of the following techniques may be used.

Placing the cache directly beside the FRP. The simplest method is to place the cache directly beside the FRP. Then pinpointing is reduced to specifying the precise reference point of the FRP (Figure 5-1).

Sighting the cache by projection. This method may be used if the FRP has one flat side long enough to permit precise sighting by projecting a line along the side of the object. The burial party places the cache a measured distance along the sighted line (Figure 5-2). This method may also be used if two precise FRPs are available, by projecting a line sighted between the two objects. In either case, the instructions for finding the cache must state the approximate direction of the cache from the FRP. Since small errors in sighting are magnified as the sighted line is extended, the cache should be placed as close to the FRP as other factors permit. Ordinarily this method becomes unreliable if the sighted line is extended beyond 50 meters.

Placing the cache at the intersection of measured lines. If two FRPs are available within several paces, the cache can be one line projected from each of the FRPs (Figure

101

Figure 5-1. A cache located directly beside the FRP.

Figure 5-2. A cache located a measured distance along a sighted line.

5-3). If this method is used, state the approximate direction of the cache from each FRP. To ensure accuracy, neither of the projected lines (from the FRPs to the point of emplacement) should be more than twice as long as the base line (between the two FRPs). If this proportion is maintained, the only limitation upon the length of the projected lines is the length of the measuring line that the recovery party is expected to carry. The recovery party should carry *two* measuring lines when this method is used.

Sighting the cache by compass azimuth. If the above methods of sighting are not feasible, one measured line may be projected by taking a compass azimuth from the FRP to the point where the cache is placed (Figure 5-4). To avoid confusion, use an azimuth to a

5-6

cardinal point of the compass (north, east, south, or west). Since compass sightings are likely to be inaccurate, a cache that is pinpointed by this method should not be placed more than 10 meters from the FRP.

Figure 5-5 explains how sighting by a compass azimuth can be combined with placing the cache at the intersection of measured lines when only one FRP is available, but a multiple cache is required. (A multiple cache is usually employed for communications equipment.) Whenever possible, use several FRPs for pinpointing a multiple cache.

Measuring Distances

The observer should express all measured distances in a linear system that the recovery party is sure to understand—ordinarily the standard system for the country where the cache is located. He should use whole numbers (6 meters, not 6.3 or 6.5) to keep his instructions as brief and as simple as possible. To get an exact location for the cache in whole numbers, take sightings and measurements first.

If the surface of the ground between the points to be measured is uneven, the linear distance should be measured on a direct line from point to point, rather than by following the contour of the ground. This method requires a measuring line long enough to reach the full distance from point to point and strong enough to be pulled taut without breaking.

Marking Techniques

The emplacement operation can be simplified and critical time saved if the point where the cache is to be buried is marked during the reconnaissance.

If a night burial is planned, the point of emplacement may have to be marked during a daylight reconnaissance. This method should be used whenever operational conditions permit.

The marker must be an object that is easily recognizable but that is meaningless to an unwitting observer. For example, a small rock or a branch with its butt placed at the point selected for the emplacement may be used.

Additional Data Required for Emplacement

During a personal reconnaissance, the observer must not only pinpoint the cache site, but also gather all the incidental information required for planning the emplacement operation. It is especially important to determine the best route to the site and at least one alternate route, the security hazards along these routes, and any information that can be used to overcome the hazards.

Since this information is also essential to the recovery operation, it must be compiled after emplacement and included in the final cache report. Therefore, the observer should be thoroughly familiar with the *Twelve-Point Cache Report* (Appendix D) before he starts a personal reconnaissance. This report is a checklist for the observer to record as much information as possible. Points 6 through 9 and 11 are particularly important. The personal reconnaissance also provides an excellent opportunity for a preliminary estimate of the time required for getting to the site.

THE ALTERNATE SITE

As a general rule, planners should select an alternate site in case unforeseen difficulties prevent use of the best site. Unless the primary site is in a completely deserted area, there is always some danger that the emplacement party will find it occupied as they approach, or that the party will be observed as they near the site. The alternate site should be far enough away to be screened from view from the primary site, but near enough so that the party can reach it without making a second trip.

THE CONCEALMENT SITE

A site that looks ideal for concealment may be revealed to the enemy for that very reason. Such a site may be equally attractive to a native of an occupied country to hide his valuables. The only real key to the ideal concealment site is careful casing of the area combined with great familiarity with local residents and their customs. The following is a list of likely concealment sites:

- Natural caves and caverns, and abandoned mines and quarries.
- Walls (hidden behind loose bricks or stones or hidden behind a plastered surface).
- Abandoned buildings.
- Infrequently used structures (stadiums and other recreational facilities, and railroad facilities on spur lines).
- Memorial edifices (mausoleums, crypts, monuments).
- Public buildings (museums, churches, libraries).
- Ruins of historical interest.
- Culverts.
- Sewers.
- Cable conduits.

The concealment site must be equally accessible to the person emplacing and the person recovering. However, visits by both persons to certain interior sites may be incompatible with the cover. For instance, a site in a house owned by a relative of the emplacer may be unsuitable because there is no adequate excuse for the recovery person to enter the house if he has no connection with the owner.

The site must remain accessible as long as the cache is needed. If access to a building depends upon a personal relationship with the owner, the death of the owner or the sale of the property might render it inaccessible.

Persons involved in the operation should not be compromised if the cache is discovered on the site. Even if a cache is completely sterile, as every cache should be, the mere fact that it has been placed in a particular site may compromise certain persons. If the cache were discovered by the police, they might suspect the emplacer because it was found in his relative's house.

The site must not be located where potentially hostile persons frequently visit. For instance, a site in a museum is not secure if police guards or curious visitors frequently enter the museum.

To preserve the cached material, the emplacer must ensure the site is physically secure for the preservation of the cached material. For example, most buildings involve a risk that the cache may be destroyed or damaged by fire, especially in wartime. The emplacer should consider all risks and weigh them against the advantages of an interior site.

5-8

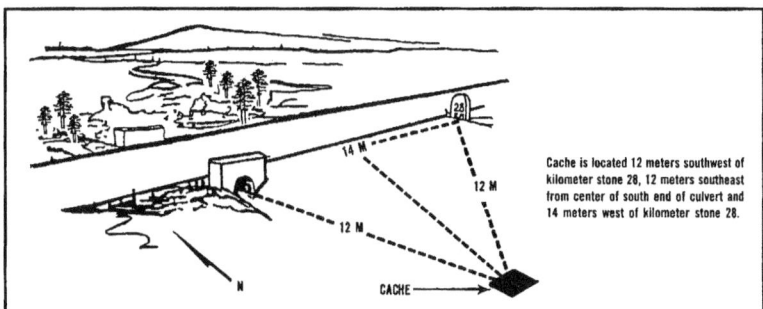

Cache is located 12 meters southwest of kilometer stone 28, 12 meters southeast from center of south end of culvert and 14 meters west of kilometer stone 28.

Figure 5-3. A cache located at the intersection of measured lines.

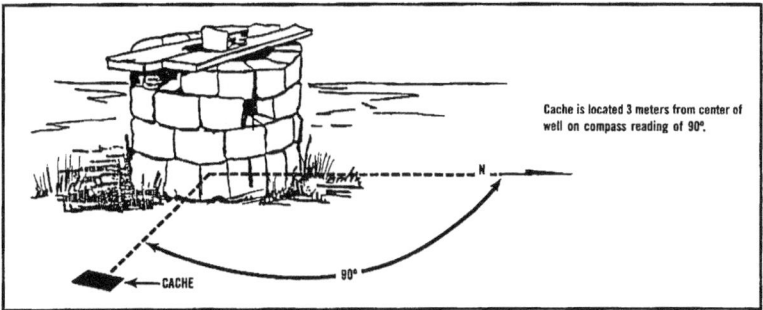

Cache is located 3 meters from center of well on compass reading of 90°.

Figure 5-4. A cache located by compass azimuth to a cardinal point.

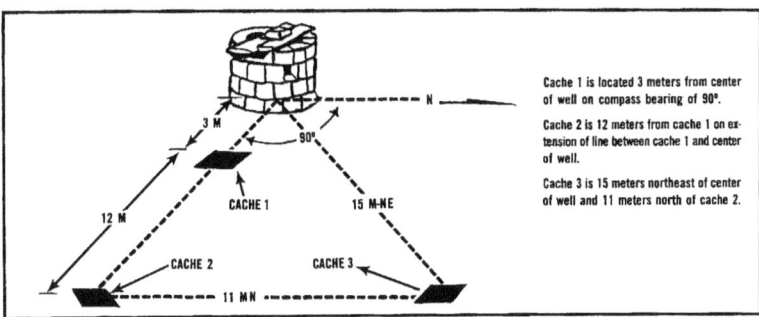

Cache 1 is located 3 meters from center of well on compass bearing of 90°.

Cache 2 is 12 meters from cache 1 on extension of line between cache 1 and center of well.

Cache 3 is 15 meters northeast of center of well and 11 meters north of cache 2.

Figure 5-5. Multiple cache located by compass azimuth and intersection of measured lines.

5-9

A custodian may serve to ease access to a building or to guard a cache. However, the use of such a person is inadvisable; as a custodian poses an additional security risk. He may use the contents of the cache for personal profit or reveal its location.

THE BURIAL SITE

In selecting a burial site, consider the following special factors along with the basic considerations of suitability and accessibility:

Drainage

This includes the elevation of the site and the type of soil. The importance of good drainage makes a site on high ground preferable unless other factors rule it out. Moisture is one of the greatest natural threats to the contents of a cache. Swamp muck is the most difficult soil to work in. If the site is near a stream or river, ensure that the cache is well above the all-year high-water mark so that it will not be uncovered if the soil is washed away.

Ground Cover

The types of vegetation at the site will influence the choice. Roots of deciduous trees make digging very difficult. Coniferous trees have less extensive root systems. Also, the presence of coniferous trees usually means that the site is well drained. Does the vegetation show paths or other indications that the site is frequented too much for secure caching? Can the ground cover be easily restored to its normal appearance when burial is completed? Tall grass reveals that it has been trampled, while an overlay of leaves and humus can be replaced easily and will effectively conceal a freshly refilled hole.

Natural Concealment

The vegetation or the surrounding terrain should offer natural concealment for the burial and recovery parties working at the site. Planners should carefully consider seasonal variations in the foliage.

Type of Soil

Sandy loam is ideal because it is easy to dig and drains well. Clay soil should be avoided because it becomes quite sticky in wet weather and in dry weather it may become so hard that it is almost impossible to dig.

Snowfall and Freezing

If the cache must be buried or recovered in winter, data on the normal snowfall, the depth to which the ground freezes in winter, and the usual dates of freezing and thawing will influence the choice of the site. Frozen ground impedes digging and requires additional time for burial and recovery. Snow on the ground is especially hazardous for the burial operation. It is practically impossible to restore the snow over the burial site to its normal appearance unless there is more snowfall or a brisk wind. Also, it is very difficult to ensure that no traces of the operation are left after the snow has melted.

Rocks and Other Subsurface Obstructions

Large obstructions that might prevent use of a particular site can be located to some extent before digging by probing with a rod or stake at the exact spot selected for the cache.

—

THE SUBMERSION SITE

To be suitable for a submerged cache, a body of water must have certain characteristics. The presence of these characteristics can be determined only by a thorough survey of the site. Their importance will be understood after familiarization with the technicalities of submersion as discussed in Section III, Emplacement. Submersion usually requires a boat, first for reconnoitering, then for emplacement. Thus, the accessibility problems involved in submersion usually narrow down to the availability of a boat and the action cover for using it. If there is no fishing or pleasure boating at the site, the cover for this peculiar type boating may be a real problem.

In tropical areas the course of streams or rivers is frequently changed by seasonal rainfall and can cause many problems. Keep this fact in mind when choosing the site and when selecting reference points.

RECOVERY

Since the method for recovering a cache is generally similar to that for emplacing (Section III) a cache, it need not be described in full. However, several important considerations should be stressed in training for a recovery operation.

Practical Exercises

Anyone who is expected to serve as a recovery person should have the experience of actually recovering dummy caches, if field exercises can be arranged securely. It is especially desirable for the recovery person to be able to master the pinpointing techniques. Mastery is best attained by practice in selecting points of emplacement and in drafting, as well as in following instructions.

Equipment

Although the equipment used in recovery is generally the same as that used in emplacement, it is important to include any additional items that may be required in recovery in the cache report. A probe rod may not be essential for emplacement, but it is necessary to have some object roughly the same size as the cache container to fill the cavity left in the ground by removal of a buried cache. Some sort of container or wrapping material may be needed to conceal the recovered cache while it is being carried from the cache site to a safe house. Recovery of a submerged cache may require grappling lines and hooks, especially if it is heavy.

Sketch of the Site

If possible, the observer should provide the recovery person with sketches of the cache site and the route to the cache site. If the recovery person must rely exclusively on verbal instructions, as is the case when communications are limited to radiotelephone (RT) messages, he should draw a sketch of the site before starting on the recovery operation. He should use all the data in the verbal instructions to make the sketch as realistic as possible. Drawing a sketch will help to clarify any misunderstanding of the instructions. Also, a sketch can be followed more easily than verbal instructions. It may also be helpful for the recovery person to draw a sketch of the route from the immediate reference point to the site. But he should not carry this sketch on him because if he were apprehended the sketch might direct the enemy to the cache.

5-11

Preliminary Reconnaissance

Checking the instructions for locating the cache may be advisable, especially when the recovery operation must be performed under stringent enemy controls or when there is no extra time for searching. Careful analysis of the best available map can minimize reconnoitering activity in the vicinity of the cache and thus reduce the danger of arousing suspicion. If recovery must be done at night, the recovery person should find the cache by daylight and place an unnoticeable marker directly over it.

Probe Rod

The recovery person can avoid digging at the wrong spot by using a probe rod before starting to dig. He should push and turn the probe rod into the ground by hand, so that it will not puncture the cache's container. Never pound the probe rod with a hammer.

Procedure for Digging and Refilling the Hole

The recovery procedure is the same as for burial, except for two points. First, never use a pick for digging the hole because it might puncture the container and damage the cached items. Second, it may be necessary to fill the hole with other objects in addition to soil after the cache is removed. Sometimes it is possible to fill the hole with rocks, sticks, or other readily available objects at the site. If no such objects are found during the preliminary reconnaissance, the recovery person should carry to the site an object roughly the same size as the cache container.

Sterilization of the Site

As with emplacement, the recovery operation must be performed in such a way that no traces of the operation are left. Although sterilizing is not as important for recovery as for emplacement, it should be done as thoroughly as time permits. Evidence that a cache has been recovered might alert the enemy to clandestine activity in the area and provoke countermeasures.

Section II. Packaging

Packaging usually involves packing the items to be cached, as well as the additional processing in protecting these items from adverse storage conditions. Proper packaging is important because inadequate packaging very likely will render the items unusable. Since special equipment and skilled technicians are needed for best results, packaging should be done at headquarters or a field packaging center whenever possible. However, to familiarize operational personnel with the fundamentals of packaging, so that they can improvise field expedients for emergency use, this section discusses determining factors, steps in packaging, wrapping materials, and criteria for the container.

DETERMINING FACTORS

The first rule of packaging is that all processing is tailored to fit the specific requirements of each cache.

The method of packaging, as well as the size, shape, and weight of the package is determined by the items to be cached, by the method of caching, and, especially, by the way the cache is recovered and used. For instance, if circumstances require one man to recover the cache by himself, the container should be no larger than a small suitcase, and the total weight of container and contents no more than 30 pounds. Of course, these limits must be exceeded with some equipment, but the need for larger packages should be weighed against the difficulties and risks in handling them. Even if more than one person is available for recovery, the material should be divided whenever possible into separate packages of a size and weight readily portable by one man.

Another very important factor in packaging concerns adverse storage conditions. Any or all of the following conditions may be present: moisture, external pressure, freezing temperatures, and the bacteria and corrosive chemicals found in some soil and water. Animal life may present a hazard; insects and rodents may attack the package. If the cache is concealed in an exterior site, larger animals also may threaten it. Whether the packaging is adequate usually depends upon how carefully the conditions at the site were analyzed in designing the cache. Thus, the method of caching (burial, concealment, or submersion) should be determined before the packaging is done.

It is equally important to consider how long the cache is to be used. Since one seldom knows when a cache will be needed, a sound rule is to design the packaging to withstand adverse storage conditions for at least as long as the normal shelf life of the contents to be cached.

STEPS IN PACKAGING

The exact procedure for packaging depends upon the specific requirements for the cache and upon the packaging equipment available. There are nine steps that are almost always necessary in packaging.

Inspecting

The items to be cached must be inspected immediately before packaging to ensure they are complete, in serviceable condition, and free of all corrosive or contaminative substances.

Cleaning

All corrodible items must be cleaned thoroughly immediately before the final preservative coating is applied. All foreign matter, including any preservative applied before the item was shipped to the field, should be removed completely. Throughout the packaging operation, all contents of the cache should be handled with rubber or freshly cleaned cotton gloves. Special handling is important because even minute particles of human sweat will corrode metallic equipment. Also, any fingerprints on the contents of the cache may enable the enemy to identify those who did the packaging.

Drying

When cleaning is completed, every trace of moisture must be removed from all corrodible items. Methods of drying include: wiping with a highly absorbent cloth, heating, or applying desiccant. Usually heating is best, unless the item can be damaged by heat. To dry by heating, the item to be cached should be placed in an oven for at least 3 hours at a temperature of about 110° F. An oven can be improvised from a large metal can or drum. In humid climate, it is especially important to dry the oven thoroughly before using it by preheating it to at least 212° F. Then, insert the equipment to be cached as soon as the oven cools down to about 110° F. If a desiccant is used, it should not touch any metallic surface. Silica gel is a satisfactory desiccant, and it is commonly available.

5-13

109

Coating With a Preservative

Apply a light coat of oil to weapons, tools, and other items with unpainted metallic surfaces. A coat of paint may suffice for other metal items.

Wrapping

When drying and coating are completed, wrap the items to be cached in a suitable material (see paragraph below on Wrapping Materials). The wrapping should be as nearly waterproof as possible. Each item should be wrapped separately, so that one perforation in the wrapping will not expose all items in the cache. The wrapping should fit tightly to each item to eliminate air pockets, and all folds should be sealed with a waterproof substance.

Packing

Several simple rules must be observed when packing items in the container. All moisture must be removed from the interior of the container by heating or applying desiccant. A long-lasting desiccant should be packed inside the container to absorb any residual moisture. If silica gel is used, the required amount can be calculated by using the ratio of 15 kilograms of silica gel to 1 cubic meter of storage space within the container. (This figure is based on two assumptions: the container is completely moistureproof and the contents are slightly moist when inserted.) Therefore, the ratio allows an ample margin for incomplete drying and can be reduced if the drying process is known to be highly effective.

Air pockets should be eliminated as much as possible by tight packing. Thoroughly dried padding should be used liberally to fill air pockets and to protect the contents from shock. Clothing and other items, which will be useful to the recovery party, should be used for padding if possible. Items made of different metals should never touch, since continued contact may cause corrosion through electrolytic action.

Enclosing Instructions for Use of Cached Equipment

Written instructions and diagrams should be included if they facilitate assembly or use of the cached items. Instructions must be written in a language that recovery personnel can understand. The wording should be as simple as possible and unmistakably clear. Diagrams should be self–explanatory since the eventual user may not be able to comprehend written instructions because of language barriers.

Sealing

When packing is completed, the lid of the container must be sealed to make it watertight. Whatever sealing device is used, it is extremely important to ensure that the sealing is done properly because the closing joint is the most vulnerable.

Testing Seal by Submersion

After the container is sealed, it should be tested to make sure that it is watertight. Testing can be done by entirely submerging the container in water and watching for escaping air bubbles. Hot water should be used if possible because hot water will bring out leaks that would not be revealed by a cold water test.

WRAPPING MATERIALS

The most important requirement for wrapping material is that it be moistureproof. Also, it should be self–sealing or adhesive to a sealing material; it should be pliable enough to fit closely, with tight folds; and it should be tough enough to resist tearing and puncturing.

Pliability and toughness may be combined by using two wrappings: an inner one that is thin and pliable and an outer one of heavier material. A tough outer wrapping is essential unless the container and the padding are adequate to prevent items from scraping together inside the cache. Five wrapping materials are recommended for field expedients because they often can be obtained locally and used effectively by unskilled personnel.

Aluminum Foil

For use as an inner wrapping, aluminum foil is the best of the widely available materials. It is moistureproof as long as it does not become perforated and provided the folds are adequately sealed. The drawbacks to its use for caching are that the thin foils perforate easily, while the heavy ones (over 2 mils thick) tend to admit moisture through the folds. The heavy-duty grade of aluminum foil generally sold for kitchen use is adequate when used with an outer wrapping. Scrim-backed foil, which is heat-sealable, is widely used commercially to package articles for shipment or storage. Portable heat-sealers that are easy to use are available commercially. Or, sealing can be done with a standard household iron.

Moisture-Resistant Papers

Several brands of commercial wrapping papers are resistant to water and grease. They do not provide lasting protection against moisture when used alone, but they are effective as an inner wrapping to prevent rubber, wax, and similar substances from sticking to the items in the cache.

Rubber Repair Gum

This is a self-sealing compound generally used for repairing tires; it makes an excellent outer wrapping. Standard commercial brands come in several thicknesses; 2 millimeters is the most satisfactory for caching. A watertight seal is produced easily by placing two rubber surfaces together and applying pressure manually. The seal should be at least 1/2 inch wide. Since rubber repair gum has a tendency to adhere to some items, an inner wrapping of nonadhesive material must be used with it, and the backing should be left on the rubber material to keep it from sticking to other items in the cache.

Grade C Barrier Material

This is a cloth impregnated with microcrystalline wax that is used extensively when packing for storage or for overseas shipment. Thus, it is generally available, and it has the additional advantage of being self-sealing. Although it is not as effective as rubber repair gum, it may be used as an outer wrapping over aluminum foil to prevent perforation of the foil. Used without an inner wrapping, three layers of *grade C barrier material* may keep the contents dry for as long as three months, but it is highly vulnerable to insects and rodents. Also, the wax wrapping has a low melting point and will adhere to many items, so it should not be used without an inner wrapping except in emergencies.

Wax Coating

If no wrapping material is available, an outer coating of microcrystalline wax, parafin, or a similar waxy substance can be used to protect the contents against moisture. It will not provide protection against insects and rodents. The package should be hot-dipped in the waxy substance, or the wax can be heated to molten form and applied with a brush.

THE CONTAINER

The outer container serves to protect the contents from shock, moisture, and other natural hazards to which the cache may be exposed.

5-15

111

Criteria for the container

The ideal container should be—

- Completely watertight and airtight after sealing.
- Noiseless when handled and its handles should not rattle against the body of the container.
- Resistant to shock and abrasion.
- Able to withstand crushing pressures.
- Lightweight in construction.
- Able to withstand rodents, insects, and bacteria.
- Equipped with a sealing device that can be closed and reopened easily and repeatedly.
- Capable of withstanding highly acidic or alkaline soil or water.

The Standard Stainless Steel Container

The standard stainless steel container comes in several sizes. Since the stainless steel container is more satisfactory than any that could be improvised in the field, it should be used whenever possible. Ideally, it should be packed at headquarters or at a field packaging center. If the items to be cached must be obtained locally, it is still advisable to use the stainless steel container because its high resistance to moisture eliminates the need for an outer wrapping. Packers should, however, use a single wrapping even with the stainless steel container to protect the contents from any residual moisture that may be present in the container when it is sealed.

The Field Expedient Container

Obviously, the ideal container cannot be improvised in the field, but the standard military and commercial containers discussed below can meet caching requirements if they are adapted with care and resourcefulness. First, a container must be sufficiently sturdy to remain unpunctured and retain its shape through whatever rough handling or crushing pressure it may encounter. (Even a slight warping may cause a joint around the lid to leak.) Second, if the lid is not already watertight and airtight, packers can make it so by improvising a sealing device. The most common type sealing device includes a rubber–composition gasket or lining and a sharp metal rim that is pressed against the gasket by a clamp or spring. The gasket must be tough and the rim must be sharp enough to indent the gasket without cutting it. Another common sealing device is a threaded lid. Its effectiveness can be increased by applying heavy grease to the threads. (Metallic solder should not be used for sealing because it corrodes metal surfaces when exposed to moisture.) Whenever any nonstainless metal container is used, it is important to apply several coats of high–quality paint to all exterior surfaces.

Instrument containers. Ordinarily, aircraft and other precision instruments are shipped in steel containers with a waterproof sealing device. The standard instrument containers range from 1/2 gallon to 10 gallon sizes. If one of suitable size can be found, only minimum modifications may be needed. In the most common type of instrument container, the only weak point is the nut and bolt that tightens the locking band around the lid. These should be replaced with a stainless steel nut and bolt.

Ammunition boxes. Several types and sizes of steel ammunition boxes that have a rubber–gasket closing device are satisfactory for buried caches. An advantage of using ammunition boxes as cache containers is that they are usually available at a military depot.

Steel drums. A caching container of suitable size may be found among the commercially used steel drums for shipping oil, grease, nails, soap, and other products. The most common types, however, lack an adequate sealing device, so a waterproof material should be used around the lid. Fully removable head drums with lock-ring closures generally give a satisfactory seal.

Glass jars. The advantage of using glass is that it is waterproof and does not allow chemicals, bacteria, and insects to pass through it. Although glass is highly vulnerable to shock, glass jars of a sturdy quality can withstand the crushing pressure normally encountered in caching. However, none of the available glass containers have an adequate sealing device for the joint around the lid. The standard commercial canning jar with a spring clamp and rubber washer is watertight, but the metal clamp is vulnerable to corrosion. Therefore, a glass jar with a spring clamp and a rubber washer is an adequate expedient for short-term caching of small items, but it should not be relied upon to resist moisture for more than a year.

Paint cans. Standard cans with reusable lids require a waterproof adhesive around the lids. It is especially important to apply several coats of paint to the exterior of standard commercial cans because the metal in these cans is not as heavy as that in metal drums. Even when the exterior is thoroughly painted, paint cans probably will not resist moisture for more than a few months.

Section III. Methods of Emplacement

Since burial is the most frequently used method of emplacement, this section describes first the complete procedure for burial, followed by a discussion of emplacement procedures peculiar to submersion and concealment. The last area discussed is the preparation of the cache report—a vital part of a caching operation.

BURIAL

When planners have designed a cache and selected the items for caching, they must carefully work out every step of the burial operation in advance.

Horizontal and Vertical Caches

Ordinarily, the hole for a buried cache is vertical (the hole is dug straight down from the surface [Figure 5-6]). Sometimes a horizontal cache, with the hole dug into the side of a steep hill or bank, provides a workable solution when a suitable site on level or slightly sloping ground is not available (Figure 5-7). A horizontal cache may provide better drainage in areas of heavy rainfall, but it is more likely to be exposed by soil erosion and more difficult to refill and restore to normal appearance.

Dimensions of the Hole

The exact dimensions of the hole, either vertical or horizontal, depend on the size and shape of the cache container. As a general rule, ensure that the hole is large enough for the container to be inserted easily. The horizontal dimensions of the hole should be about 30 centimeters longer and wider than the container. Most importantly, it should be deep enough to permit covering the container with soil to about 45 centimeters. This figure is recommended for normal usage because a more shallow burial risks exposure of the cache through soil erosion or inadvertent uncovering by normal indigenous activity. A deeper hole makes probing for recovery more difficult and unnecessarily prolongs the time required for burial and recovery.

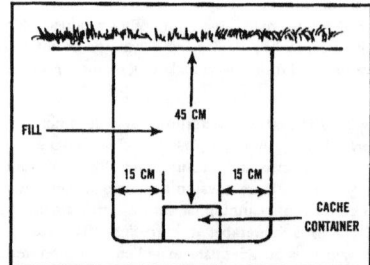

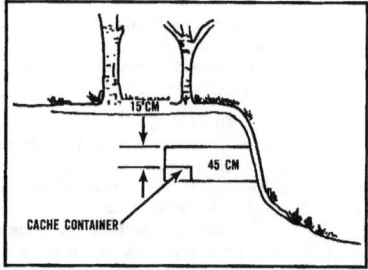

Figure 5-6. Vertical hole for a buried cache. Figure 5-7. Horizontal hole for a buried cache.

Excavation Shoring

If there is a risk that the surrounding soil will cave in during excavation, boards or bags filled with subsoil may be used to shore the sides of the hole. Permanent shoring may be needed to protect an improvised container from pressure or shock.

Equipment

The following items of equipment may be helpful or indispensable in burying a cache, depending upon the conditions at the site:

- Measuring instruments (a wire or metal tape and compass) for pinpointing the site.
- Paper and pencil for recording the measurements.
- A probe rod for locating rocks, large roots, or other obstacles in the subsoil (Figure 5-8).
- Two ground sheets on which to place sod and loose soil. An article of clothing may be used for a small excavation if nothing else is available.
- Sacks (sandbags, flour sacks) for holding subsoil.
- A spade or pickax, if the ground is too hard for spading.
- A hatchet for cutting roots.

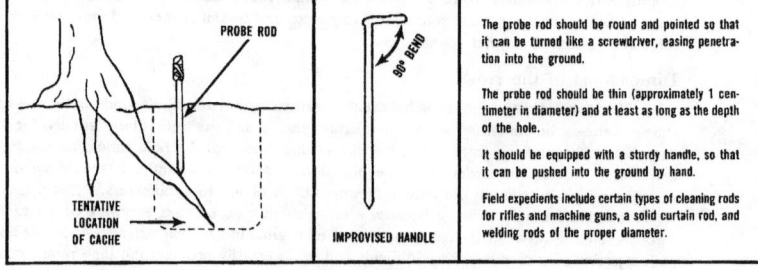

Figure 5-8. Using a probe rod.

- A crowbar for prying rocks.
- A flashlight or lamp if burial is to be done at night.

The Burial Party

Aside from locating, digging, and refilling the hole, the most important factors in this part of the emplacement operation may be expressed with one word: *Personnel*. Since it is almost impossible to prevent every member of the burial party from knowing the location of the cache, each member is a prime security problem as long as the cache remains intact. Thus, planners must keep the burial party as small as possible and select each member with utmost care. Once selected, each member must have adequate cover to explain his absence from home or work during the operation, his trip to and from the site, and his possession of whatever equipment that cannot be concealed on the way. Transportation for the burial party may be a problem, depending on the number of persons, how far they must go, and what equipment they must take. When planners have worked out all details of the operation, they must brief every member of the burial party on exactly what he is to do from start to finish.

The Operational Schedule

The final step in planning the emplacement operation is to make a schedule to set the date, time, and place for every step of the operation that requires advance coordination. The schedule will depend mainly on the circumstances, but to be practical it must include a realistic estimate of how long it will take to complete the burial. Here generalizations are worthless, and the only sure guide is actual experience under similar conditions. Planners should consider three things with respect to scheduling.

A careful burial job probably will take longer than most novices will expect. Therefore, if circumstances require a tight schedule, a *dry run* or test exercise before taking the package to the site may be advisable.

Unless, the site is exceptionally well concealed or isolated, night burial probably will be required to avoid detection. Because of the difficulties of working in the dark, a nighttime practice exercise is especially advisable.

The schedule should permit waiting for advantageous weather conditions. The difficulties of snow have already been mentioned. Rainy weather increases the problems of digging and complicates the cover story. If the burial is to be done at night, a moonless or a heavy overcast night is desirable.

Site Approach

Regardless of how effective the cover of actions during the trip to the cache site, the immediate approach must be completely unobserved to avoid detection of the burial. To reduce the risk of the party being observed, planners must carefully select the point where the burial party *disappears*, perhaps by turning off a road into woods. They should as carefully select the *reappearance* point. In addition, the return trip should be by a different route. The burial party should strictly observe the rule for concealed movement. The party should proceed cautiously and silently along a route that makes the best use of natural concealment. Concealed movement requires foresight, with special attention to using natural concealment while reconnoitering the route and to preventing rattles when preparing the package and contents.

5-19

Security Measures at the Site

The burial party must maintain maximum vigilance at the cache site, since detection can be disastrous. The time spent at the site is the most critical.

At least one lookout should be on guard constantly. If one man must do the burial by himself, he should pause frequently to look and listen.

The burial party should use flashlights or lanterns as little as possible, and should take special care to mask the glare.

Planning should include emergency actions in case the burial party is interrupted. The party should be so thoroughly briefed that it will respond instantly to any sign of danger.

Planners should also consider the various escape routes and whether the party will attempt to retain the package or conceal it along the escape route.

Steps in Digging and Refilling

Although procedures will vary slightly with the design of the cache, persons involved in caching operations must never overlook certain basic steps. The whole procedure is designed to restore the site to normal as far as possible as shown in Figure 5-9.

Site Sterilization

When the hole is refilled, make a special effort to ensure that the site is left sterile—restored to normal in every way, with no clues left to indicate burial or the burial party's visit to the vicinity. Since sterilization is most important for the security of the operation, the schedule should allow ample time to complete these final steps in an unhurried, thorough manner.

Dispose of any excess soil far enough away from the site to avoid attracting attention to the site. Flushing the excess soil into a stream is the ideal solution.

Check all tools and equipment against a checklist to ensure that nothing is left behind. This should include all personal items that may drop from pockets. To keep this risk to a minimum, members of the burial party should carry nothing on their persons except the essentials for doing the job and covering their actions.

Make a final inspection of the site for any traces of the burial. Because this step is more difficult on a dark night, use of a carefully prepared checklist is essential. With a night burial, returning to the site in the daytime to inspect it for telltale evidence may be advisable, if this can be done safely.

SUBMERSION

Emplacing a submerged cache always involves two basic steps: weighting the container to keep it from floating to the surface and mooring it to keep it in place.

Container Weighting and Mooring

Ordinarily, container weights rest on the bottom of the lake or river and function as anchors, and the moorings connect the anchors to the container. The moorings must also serve a second function, that is to provide a handle for pulling the cache to the surface when it is recovered. If the moorings are not accessible for recovery, another line must extend from the cache to a fixed, accessible object in the water or on shore. There are four types of moorings.

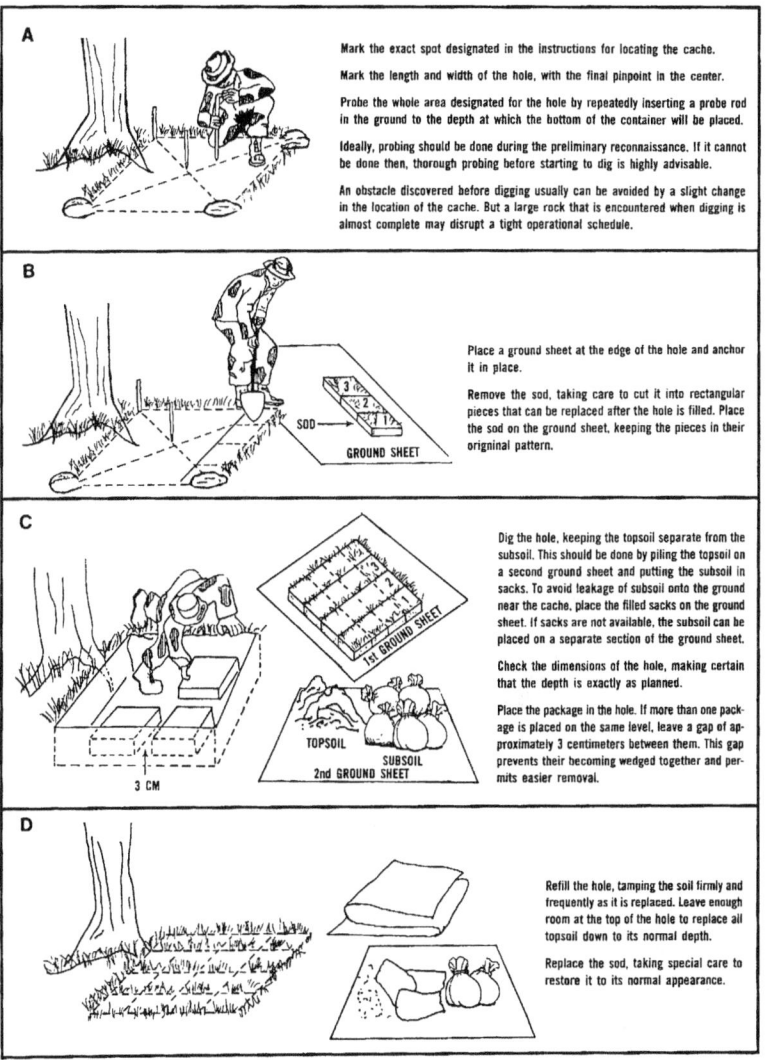

A

Mark the exact spot designated in the instructions for locating the cache.

Mark the length and width of the hole, with the final pinpoint in the center.

Probe the whole area designated for the hole by repeatedly inserting a probe rod in the ground to the depth at which the bottom of the container will be placed.

Ideally, probing should be done during the preliminary reconnaissance. If it cannot be done then, thorough probing before starting to dig is highly advisable.

An obstacle discovered before digging usually can be avoided by a slight change in the location of the cache. But a large rock that is encountered when digging is almost complete may disrupt a tight operational schedule.

B

Place a ground sheet at the edge of the hole and anchor it in place.

Remove the sod, taking care to cut it into rectangular pieces that can be replaced after the hole is filled. Place the sod on the ground sheet, keeping the pieces in their orignial pattern.

C

Dig the hole, keeping the topsoil separate from the subsoil. This should be done by piling the topsoil on a second ground sheet and putting the subsoil in sacks. To avoid leakage of subsoil onto the ground near the cache, place the filled sacks on the ground sheet. If sacks are not available, the subsoil can be placed on a separate section of the ground sheet.

Check the dimensions of the hole, making certain that the depth is exactly as planned.

Place the package in the hole. If more than one package is placed on the same level, leave a gap of approximately 3 centimeters between them. This gap prevents their becoming wedged together and permits easier removal.

D

Refill the hole, tamping the soil firmly and frequently as it is replaced. Leave enough room at the top of the hole to replace all topsoil down to its normal depth.

Replace the sod, taking special care to restore it to its normal appearance.

Figure 5-9. Digging and refilling the hole.

5-21

117

Spider Web Mooring. The container is attached to several mooring cables that radiate to anchors placed around it to form a web. The container must be buoyant so that it lifts the cables far enough off the bottom to be readily secured by grappling. The site must be located exactly at the time of emplacement by visual sightings to fixed landmarks in the water or along the shore, using several FRPs to establish a point where two sighted lines intersect. For example, in Figure 5–10, the cache is located in line with the south side of the pier and on the extension line between the east side of the spillway and chimney of the paper mill, south of the pond. For recovery, the site is located by taking sightings on the reference points, then a mooring cable is engaged by dragging the bottom or diving. This method of mooring is most difficult for recovery. It can be used only where the bottom is smooth and firm enough for dragging, or where the water is not too deep, cold, or murky for diving.

Line-to-shore mooring. A line is run from the weighted container to an immovable object along the shore (Figure 5–11). The section of the line that extends from the shore to the container must be buried in the ground or otherwise well concealed.

Buoy mooring. A line is run from the weighted container to a buoy or other fixed, floating marker, and fastened well below the waterline (Figure 5–12). This method is secure only as long as the buoy is left in place. Buoys are generally inspected and repainted every six months or so. The inspection schedule should be determined before a buoy is used.

Structural mooring. A line for retrieving the weighted container is run to a bridge pier or other solid structure in the water. This line must be fastened well below the low–water mark. In Figure 5–11, the cache is moored to the fifth piling from the west end of the pier on the south side.

Essential Data for Submersion

Whatever method of mooring is used, planners must carefully consider certain data before designing a submerged cache. The cache very likely will be lost if any of the following critical factors are overlooked:

Buoyancy. Many containers are buoyant even when filled, so the container must be weighted sufficiently to submerge it and keep it in place. If the contents do not provide enough weight, emplacers must make up the balance by attaching a weight to the container. The approximate weight needed to attain zero buoyancy is shown in Figure 5–14. This figure applies to several sizes of stainless steel containers.

Cache is located in line with south side of pier and on extension line between east side of spillway and chimney of papermill, south of pond.

SPIDERWEB MOORING

CACHE

Figure 5-10. Spiderweb mooring.

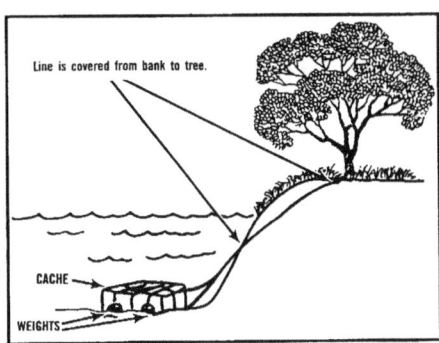

Figure 5-11. Line-to-shore mooring.

Figure 5-12. Buoy mooring.

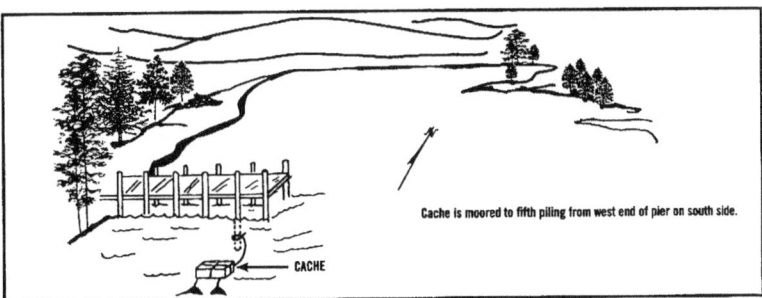

Figure 5-13. Structural mooring.

The weighting required for any container can be calculated theoretically if the displacement of the container and the gross weight of the container plus its contents are known. This calculation may be useful for designing an anchor, but it should not be relied upon for actual emplacement. To avoid hurried improvisation during emplacement, emplacers should always test the buoyancy in advance by actually submerging the weighted container. This test determines only that a submerged cache will not float to the surface. Additional weighting may be required to keep it from drifting along the bottom. As a general rule, the additional weight should be at least one-tenth of the gross weight required to make the container sink; more weight is advisable if strong currents are present.

Submersion depth. Planners must first determine the depth which the container is to be submerged to calculate the water pressure that the container must withstand. The greater the depth, the greater the danger that the container will be crushed by water pressure. For instance, the standard stainless steel burial container will buckle at a depth of approximately 4.3 meters. The difficulty of waterproofing also increases with depth. Thus, the container should not be submerged any deeper than necessary to avoid detection. As a general rule, 2.2 meters is the maximum advisable depth for caching. If seasonal or tidal variations in the

Container dimensions (inches)	Empty container weight (pounds)	Approximate weight that must be added to empty container weight to attain zero buoyancy (pounds)
7 X 9 X 8½	5	15
7 X 9 X 16½	8	31
7 X 9 X 40	16	77
7 X 9 X 45	17½	88
7 X 9 X 50	19	97

Figure 5-14. Stainless steel container buoyancy requirements.

water level require deeper submersion, the container should be tested by actual submersion to the maximum depth it must withstand.

Depth of the water. Emplacers must measure accurately the depth of the water at the point where the cache is to be placed. This will be the submersion depth if the cache is designed so that the container rests on the bottom of the lake or river. The container may be suspended some distance above the bottom, but the depth of the water must be known to determine the length of moorings connecting the containers to the anchors.

High- and low-water marks. Any tidal or seasonal changes in the depth of the water should be estimated as accurately as possible. Emplacers must consider the low-water mark to ensure that low water will not leave the cache exposed. The high-water point also should be considered to ensure that the increased depth will not crush the container or prevent recovery.

Type of bottom. Emplacers should probe as thoroughly as possible the bed of the lake or river in the vicinity of the cache. If the bottom is soft and silty, the cache may sink into the muck, become covered with sediment, or drift out of place. If the bottom is rocky or covered with debris, the moorings may become snagged. Any of these conditions may make recovery very difficult.

Water motion. Emplacers should consider tides, currents, and waves because any water motion will put additional strain on the moorings of the cache. Moorings must be strong enough to withstand the greatest possible strain. If the water motion tends to rock the cache, emplacers must take special care to prevent the moorings from rubbing and fraying.

Clearness of the water. When deciding how deep to submerge the cache, emplacers must first determine how far the cache can be seen through the water. If the water is clear, the cache may need to be camouflaged by painting the container to match the bottom. (Always paint shiny metallic fixtures a dull color.) Very murky water makes recovery by divers more difficult.

Water temperature. Planners must consider seasonal changes in the temperature of the water. Recovery may be impossible in the winter if the water freezes. The dates when the lake or river usually freezes and thaws should be determined as accurately as possible.

Salt water. Since seawater is much more corrosive than fresh water, tidal estuaries and lagoons should not be used for caching. The only exception is the maritime resupply operation, where equipment may be submerged temporarily along the seacoast until it can be recovered by a shore party.

CONCEALMENT

There are many different ways to conceal a cache in natural or ready-made hiding places. For instance, if a caching party were hiding weapons and ammunition in a cave, relying entirely on natural concealment, the emplacement operation would be reduced to simply locating the site. No tools would be needed except paper, pencil, and a flashlight. On the other hand, if the party were sealing a packet of jewels in a brick wall, a skilled mason would be needed, his kit of tools, and a supply of mortar expertly mixed to match the original brick wall.

When planning for concealment, planners must know the local residents and their customs. During the actual emplacement, the caching party must ensure the operation is not observed. The final sterilization of the site is especially important, since a concealment site is usually open to frequent observation.

CACHING COMMUNICATIONS EQUIPMENT

As a general rule, all equipment for a particular purpose (demolitions, survival) should be included in one container. Some equipment, however, is so sensitive from a security standpoint that it should be packed in several containers and cached in different locations to minimize the danger of discovery by the enemy. This is particularly true of communications equipment, since under some circumstances anyone who acquires a whole RT set with a signal plan and cryptographic material would be able to play the set back. An especially dangerous type of penetration would result. In the face of this danger, the signal plan and the cryptographic material must never be placed in the same container. Ideally a communications kit should be distributed among three containers and cached in different locations. If three containers are used, the distribution may be as follows:

- Container #1: The RT set, including the crystals.

- Container #2: The signal plan and operational supplies for the RT operator, such as currency, barter items, and small arms.

- Container #3: The cryptographic material.

When several containers are used for one set of equipment, they must be placed far enough apart so that if one is discovered, the others will not be detected in the immediate vicinity. On the other hand, they should be located close enough together so that they can be recovered conveniently in one operation. The distance between containers will depend on the particular situation, but ordinarily they should be at least 10 meters apart. One final reference point ordinarily is used for a multiple cache. (See Figure 5-4, which illustrates the use of one round FRP and a compass azimuth to pinpoint a multiple cache, and Figure 5-5, which shows how three corners on a rectangular FRP can pinpoint a multiple cache without using a compass azimuth.) The caching party should be careful to avoid placing multiple caches in a repeated pattern. Discovery of one multiple cache would give the opposition a guide for probing others placed in a similar pattern.

CACHING MEDICAL EQUIPMENT

A feasibility study must be performed to determine the need for the caching of medical supplies. The purpose of caches is to store excess medical supplies, to maintain mobility, and deny access to the enemy. Also caching large stockpiles of medical supplies allows prepositioning vital supplies in anticipation of future planned operations.

THE CACHE REPORT

The final step, which is vital in every emplacement operation, is the preparation of a cache report. This report records the essential data for recovery. The cache report must provide all the information that someone unfamiliar with the locality needs to find his way to the site, recover the cache, and return safely.

Content

The most important parts of the cache report must include instructions for finding and recovering the cache. It should also include any other information that will ease planning the recovery operation. Since the details will depend upon the situation and the particular needs of each organization, the exact format of the report cannot be prescribed. The Twelve-Point Cache Report is intended merely to point out the minimum essential data (see Appendix D, Twelve-Point Cache Report Format). Whatever format is used, the importance of attention to detail cannot be overemphasized. A careless error or omission in the cache report may prevent recovery of the cache when it is needed.

Procedure

The observer should collect as much data as possible during the personal reconnaissance to assist in selecting a site and planning emplacement and recovery operations. Drafting the cache report before emplacement is also advisable. Following these procedures will reveal the omissions. Then the missing data can be obtained at the site. If this procedure is followed, the preparation of the final cache report will be reduced to an after-action check. This check ensures that the cache actually was placed precisely where planned and that all other descriptive details are accurate. Although this ideal may seldom be realized, two procedures always should be followed:

- The caching party should complete the final cache report as soon as possible after emplacement, as details are fresh in mind.

- Someone who has not visited the site should check the instructions by using them to lead the party to the site. When no such person is available, someone should visit the site shortly after emplacement, provided he can do so securely. If the cache has been emplaced at night, a visit to the site in daylight may also provide an opportunity to check on the sterilization of the site.

5-26

CHAPTER 6

COMBAT OPERATIONS IN AN URBAN ENVIRONMENT

The Army expects urban combat to be more frequent than ever before. According to US Army doctrine, urban combat is classified as a special operation. A military operation on urban terrain is special only because it requires unique tactics and techniques. For a more detailed discussion of combat operations in an urban environment, see FM 90-10-1.

Section I. Fighting in Built-Up Areas

Successful combat operations in urban areas require unique combat skills. This section discusses some of those skills.

MOVING

Movement in urban areas is a fundamental skill that you, as an SF soldier, must master. To minimize exposure to enemy fire while moving—

- Do not silhouette yourself, stay low, avoid open areas, such as streets, alleys, and parks.
- Select your next covered position before moving.
- Conceal movements by using smoke, buildings, rubble, or foliage.
- Move rapidly from one position to another.
- Do not mask your overwatching or covering fire when moving; stay alert and ready.

CROSSING A WALL

Always cross a wall rapidly. But first, find a low spot to cross and visually reconnoiter the other side of the wall to see if it is clear of obstacles and the enemy. Next, quickly roll over the wall, keeping a low silhouette (Figure 6-1). The rapid movement and low silhouette keep the enemy from getting a good shot at you.

MOVING AROUND A CORNER

Before moving around a corner, check out the area beyond it to see if it is clear of obstacles and the enemy. Do not expose yourself when checking out the area. Lie flat on the ground and do not expose your weapon beyond the corner. Look around the corner at ground level only enough to see around it (Figure 6-2). Do not expose your head any more than necessary. If there are no obstacles or enemy present, stay low and move around the corner.

MOVING PAST A WINDOW

When moving past a window on the first floor of a building, stay below the window level (Figure 6-3). Take care not to silhouette yourself in the window, and stay close to the side of the building.

When moving past a window in a basement, use the same basic techniques used in passing a window on the first floor. However, instead of staying below the window, step or jump over it without exposing your legs (Figure 6-4).

Figure 6-1. Crossing a wall.

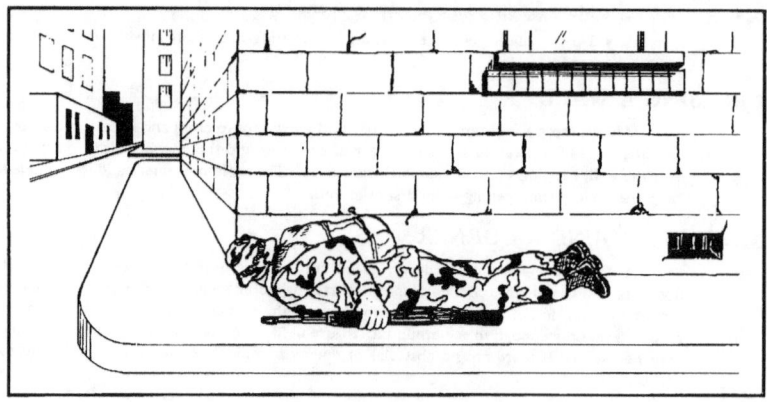

Figure 6-2. Corner movement.

6-2

124

MOVING PARALLEL TO A BUILDING

When moving parallel to a building, use smoke for concealment and have someone to overwatch your move. Stay close to the side of the building (Figure 6-5). Use shadows if possible, and stay low. Move quickly from covered position to covered position.

CROSSING OPEN AREAS

Whenever possible, avoid kill zones such as streets, alleys, and parks. They are natural kill zones for enemy machine guns. When you must cross an open area, do it quickly. Use the shortest route across the area. Use smoke to conceal your move and have someone overwatch you.

If you must go from point A to point C, as depicted in Figure 6-6, do not move from point A straight to point C. This is the longest route across the open area and gives the enemy more time to track and hit you.

Instead of going from point A straight to point C, select a place (point B) to move to, using the shortest route across the open area.

Once on the other side of the open area, move to point C using the techniques already discussed.

MOVING IN A BUILDING

When moving in a building, do not silhouette yourself in doorways and windows. Move past them as discussed for outside movement (Figure 6-7).

If forced to use a hallway, do not present a large target to the enemy. Hug the wall and get out of the hallway quickly (Figure 6-8).

ENTERING A BUILDING

When entering a building, take precautions to get into it with minimum exposure to enemy fire and observation.

Figure 6-3. High window movement.

Figure 6-4. Basement window movement.

6-3

Basic Rules

Some basic rules are the following:

- Select an entry point before moving.
- Avoid windows and doors.
- Use smoke for concealment.
- Make new entry points by using demolitions or tank rounds.
- Throw a hand grenade through the entry point before entering.
- Quickly follow the explosion of the hand grenade.
- Have your buddy overwatch you as you enter the building.
- Enter at the highest level possible.

Figure 6–5. Parallel movement.

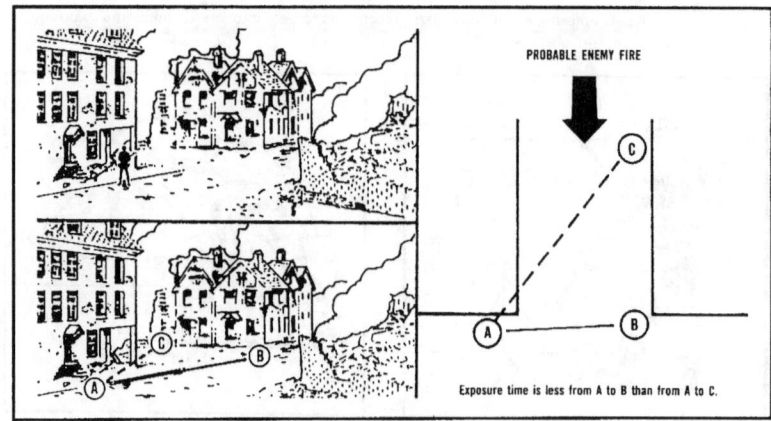

Figure 6–6. Route selection.

6–4

Figure 6-7. Window movement.

Figure 6-8. Hallway movement.

6-5

127

High-Level Entries

The preferred way to clear a building is to clear from the top down. Therefore, you should enter at the highest level possible. If the enemy is forced down to the ground level, it may leave the building, thus exposing itself to the fires outside the building (Figure 6-9).

If the enemy is forced up to the top floor, it may fight even harder than normal or escape over roofs of other buildings.

You can use ropes, ladders, drainpipes, vines, helicopters, or the roofs and windows of adjoining buildings to reach the top floor or roof of another building (Figure 6-10). You can climb onto a buddy's shoulders and pull yourself up. Or, you can attach a grappling hook to one end of a rope and throw the hook to the roof, where it can snag something to hold the rope in place.

Low-Level Entries

There will be times when you cannot enter from an upper level or the roof. In such cases, you may have to enter at the ground floor. When making low-level entries, avoid using windows and doors as much as possible. They are often booby-trapped and are probably covered by enemy fire.

When making low-level entries, use demolitions, artillery, tanks, antitank weapons, or similar means to make an entry point in a wall. Before entering the entry point, throw a cooked-off hand grenade through the entry point to reinforce the effects of the first blast (Figure 6-11).

USING HAND GRENADES

When fighting in built-up areas, use hand grenades to clear rooms, hallways, and buildings. Throw a hand grenade before entering a door, window, room, hall, stairwell, or any other entry point. Before throwing a hand grenade, let it cook off for 2 seconds. The cookoff keeps the enemy from throwing it back before it explodes.

Figure 6-9. High-level entry.

Figure 6–10. Roof landing.

Figure 6–11. Low–level entry.

6–7

129

To cook off a hand grenade, remove your thumb from the safety lever; allow the lever to rotate out and away from the grenade; then count one thousand one, one thousand two, and throw it.

The best way to put a grenade into an upper–story opening is to use a grenade launcher.

When you throw a hand grenade into an opening, stay close to the building, using it for cover (Figure 6–12). Before you throw the hand grenade, select a safe place to move to in case the hand grenade does not go into the opening or in case the enemy throws it back. Once you throw the hand grenade, take cover. After the hand grenade explodes, move into the building quickly.

USING FIGHTING POSITIONS

Fighting positions in urban areas are different from those in other types of terrain. In some cases, you must use hasty fighting positions that are no more than whatever cover is available.

Corners of Buildings

When using a corner of a building as a fighting position, you must be able to fire from either shoulder. Firing from the shoulder lets you keep your body close to the wall of the building and exposes as little of yourself as possible. If possible, fire from the prone position (Figure 6–13).

Walls

When firing from behind a wall, fire around it if possible, not over it (Figure 6–14). Firing around it reduces the chance of being seen by the enemy. Always stay low, close to the wall, and fire from the shoulder that lets you stay behind cover.

Windows

When using a window as a fighting position, do not use a standing position, as it exposes most of your body. Standing may also silhouette you against a light–colored interior

Figure 6–12. Hand grenades.

Figure 6–13. Corner position.

6–8

wall or a window on the other side of the building. Do not let the muzzle of your rifle extend beyond the window, as that may give away your position. The enemy may see the muzzle or the flash of the rifle.

The best way to fire from a window is to get well back into the room (Figure 6–15). This way prevents the muzzle or flash from being seen. Kneel to reduce exposure.

To improve the cover provided by a window, barricade the window but leave a small hole to fire through. Also barricade the other windows around your position to keep the enemy from knowing which window is being used for the fighting position. Use boards from the interior walls of the building or any other material to barricade the windows. This material should be put on in an irregular pattern. Place sandbags below and on the sides of the window that is to be used as a fighting position to reinforce it and to add cover. Remove all the glass in the window to prevent injury from flying glass (Figure 6–16).

Peaks of Roofs

A peak of a roof can provide a vantage point and cover for a fighting position (Figure 6–17). It is especially good for a sniper position. When firing from a rooftop, stay low and do not silhouette yourself.

A chimney, smokestack, or any other structure extending from a roof can provide a base behind which you can prepare a position. If possible, remove some of the roofing material so that you can stand inside the building on a beam or platform with only your head and shoulders above the roof. Use sandbags to provide extra cover (Figure 6–18).

If there are no structures extending from the roof, prepare the position from underneath the roof and on the enemy side. Remove enough of the roofing material to let

Figure 6–14. Wall position.

Figure 6-15. Window position stance.

Figure 6-16. Reinforced window position.

Figure 6-17. Rooftop position.

Figure 6-18. Position inside roof.

6-10

you see and cover your sector through it. Use sandbags to add cover. Stand back from the opening and do not let the muzzle or flash of your rifle show through the hole. The only thing that should be noticeable to the enemy is the missing roofing material (Figure 6-19).

Loopholes

A loophole blown or cut in a wall provides cover for a fighting position. Loopholes reduce the number of windows that have to be used. Cut or blow several loopholes in a wall so the enemy cannot tell which one you are using. When using a loophole, stay back from it. Do not let the muzzle or flash of your rifle show through it (Figure 6-20).

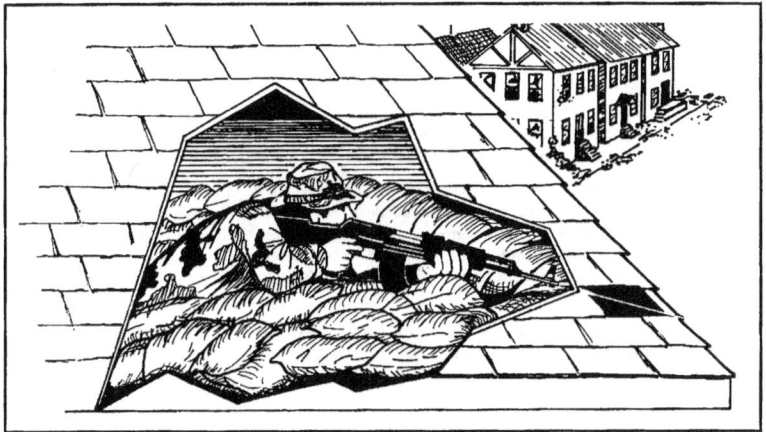

Figure 6-19. Position beneath roof.

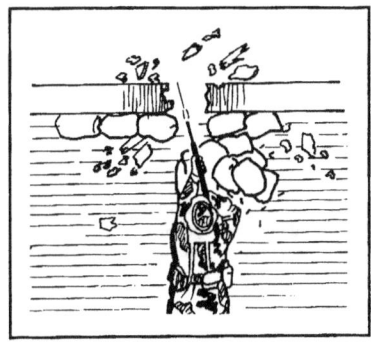

Figure 6-20. Loophole.

Figure 6-21. Reinforced loophole.

6-11

To reinforce a loophole and add cover, put sandbags around it. If firing from a prone position on the second floor, put sandbags on the floor to lie on. The sandbags will protect you from explosions on the first floor. Use a table with sandbags on it or some other sturdy structure to provide overhead cover. This cover will protect you from falling debris (Figure 6-21).

Section II. Assault Tactics in a Counterterrorist Role

This section addresses assault tactics of the special reaction team (SRT) operating in a special threat hostage situation. The SRT assault phase in a special threat hostage situation is the most difficult. It is emphasized because it poses the most serious threat to innocent bystanders. The SRT may modify assault tactics to counter operations by snipers, barricaded criminals, or terrorists.

THE SPECIAL REACTION TEAM

All special threats are high-risk situations, and pose a grave danger to hostages, bystanders, and law enforcement personnel. The extent of the danger depends in great part upon the actions and training of the personnel used to neutralize the threat.

Because locations, motivations, response capabilities, and other vital circumstances differ, special threat situations must be dealt with on an individual basis. However, to maintain the safety of all concerned, to apprehend or neutralize the offender, and to gain release of the hostages unharmed, clear and decisive coordinated threat management actions are required. Such actions must be geared to the situation and be flexible in order to meet unforeseen developments.

The safety of hostages and their eventual release without injury must be the basis for plans and actions taken during a special threat situation, except when nuclear weapons are involved. In this special situation, weapons recovery is the overriding consideration (AR 50-5).

The mission of the SRT in order of priority is—

- Protection of lives (hostages, law enforcement personnel, bystanders, and suspects).
- Safe release of hostage(s).
- Apprehension of the offender.
- Isolation of the incident.
- Protection of property and equipment.
- Conduct of the assault.

ASSAULT CONSIDERATIONS

This section and FM 19-10 outline the composition of the SRT and provide for varying team sizes. Also discussed is the composition, duties, and employment aspects of the threat management force. Field Manual 19-10 is not binding, and commanders are afforded the latitude to alter SRT size, duties, and equipment as they deem necessary.

The hostage situation is normally resolved by negotiation, wherein some part of the terrorists' demands may be met in exchange for the release of the hostages. In the extreme, this may even include allowing the offender to depart the immediate area with his hostages, although this is to be avoided, if possible.

RECONNAISSANCE AND SURVEILLANCE	Perform ground-air reconnaissance of the site as needed. Make maximum use of photography. Include SRT members (preferably the SRT scout or team leader) in reconnaissance efforts, especially ground reconnaissance.
INTERROGATION AND DEBRIEFING	Conduct interrogation and debriefing of all released hostages and friends and relatives of hostages and barricaded suspects.
RECORDS	Examine available personnel, medical, school and criminal records of the hostages and offenders. Authorize liaison with other agencies or installations to effect record examinations.
DOCUMENTS	Analyze all notes and messages obtained from the offenders. Make all information acquired immediately available to concerned personnel and agencies (civilian and military).
TECHNICAL INTELLIGENCE	Inspect all tangible items associated with the situation. Take care not to destroy the evidence value of the inspected material. Mark and catalog evidence, and establish a chain of custody to ensure its safekeeping.
MAPS, PHOTOGRAPHS, AND TERRAIN MODELS	Maintain a small-scale map of the hostage site, supplemented by photos of the area. Include floor plans of the hostage site and adjacent buildings. If time permits, develop a three-dimensional model of the site.
WEATHER	Keep up-to-date weather conditions and forecasts posted in a common area, accessible to concerned personnel and agencies.
COMMUNICATIONS	Establish communications with inner-perimeter personnel.
ALL SOURCE INTELLIGENCE	Do not limit intelligence gathering. Exploit additional sources of criminal information and intelligence as appropriate. Use established criminal information and intelligence rules to ensure that the prosecution is not compromised.
CLANDESTINE LISTENING DEVICES	Use clandestine listening devices if procedures for emergency nonconsensual interception are strictly followed (AR 190-53, paragraph 2-3).
COUNTER-INTELLIGENCE	Coordinate proposed counterintelligence measures with the emergency operations center (EOC) staff. These measures must be approved by the commander and be in accordance with AR 380-13.

Figure 6-22. Information gathering methods.

6-13

135

OPERATIONAL CONSIDERATIONS

An assault by the SRT is a last resort action, taken only when it is recognized that the offenders intend to harm the hostages. Negotiators may have failed, and no other force option is available.

To help the SRT, if an assault is necessary, acquire as much information as possible during the negotiation phase using the sources and information-gathering methods shown in Figure 6-22.

PRIORITY INTELLIGENCE REQUIREMENTS

Before undertaking a mission, make every attempt to determine the priority intelligence requirements by answering the questions shown in Figure 6-23.

PREPARATION

Speed of response is important in nearly all SRT missions, but speed is secondary to preparation. The SRT's planning must be continuous, meticulous, flexible, and detailed, and should be accomplished in accordance with the principles outlined in FM 19-10. Planning should emphasize positive command and control. The split-second timing actions among the elements of the SRT is absolutely essential. This fine timing is honed during continuous rehearsals under realistic conditions before employing the SRT. During the assault, this timing is maintained through close radio coordination. The SRT planning phase starts as soon as the hostage situation develops and continues until the situation is resolved or the SRT is committed. The SRT must always be prepared to be committed with virtually no notice when a hostage situation develops. Viable contingency plans are required. Circumstances may condense preparation time to an hour or less.

The SRT leader should determine the logistical requirements needed to complete the mission. Figure 6-24 illustrates equipment needed en route to the objective, equipment and weapons needed at the objective, and equipment needed for control during the mission. The leader should determine whether carrying special equipment or weapons would jeopardize infiltration security. He should also determine whether additional firepower and special equipment is worth the loss of speed, mobility, and deception. Weapons and equipment selections depend on information gathered concerning the type of building to be entered, the suspects and their location, and the weapons they are using.

Medical contingencies must be anticipated to address not only SRT casualties, but also captured offenders and recovered hostages. Immediate first aid, evacuation, hospitalization, and more definitive treatment must be pre-planned. Having a standby surgical team *on-site* may be reasonable, and additional security measures to protect this team may be required.

INFILTRATION TECHNIQUES

The SRT, when mobilized, should move to an assembly area sufficiently distant from the hostage site to avoid detection. Its purpose is to seize, on order, a barricaded objective that is defended by offenders. It engages the offenders by fire, as the situation dictates, captures or kills them, and frees the hostages. Figure 6-25 lists some rules for infiltration.

During the movement from the assembly area to secure the inner perimeter, cover and concealment are vital. The SRT should move to the inner perimeter, completely undetected. Even though the location of the offenders is generally known, the SRT members should be continually alert to the possibility of being fired upon from other positions.

HOSTAGES	What is the detailed description of the hostages?
	    • How many hostages are there?
	    • Who are they and what are their names?
	    • What is their native language?
	    • What is their mental and physical state?
	    • What are they wearing?
	What is their exact physical location?
	What supplies are available to them?

OFFENDERS	What is the detailed description of the offenders?
	    • How many offenders are there?
	    • Who are they and what are their names?
	    • What is their native language?
	    • What is their mental and physical state?
	    • What are they wearing?
	    • How are they armed?
	What are their intentions, motivations, and demands?
	What are their strengths and weakness?
	To what organization do they belong (military, social, civilian)?
	Are there any present or former members of this organization available?
	What methods of operation have been used by this organization in the past?
	What are the short-term objectives of the organization?
	Does the organization support the actions of its members and do the members support the goal of the organization?
	Are there any local, national, or international personnel or organizations sympathetic to these actions?
	How are they equipped (available weapons, ammunition, demolitions, and protective equipment)?
	What means of communications are available to them (telephones, radios, television, shortwave receivers and transmitters, monitoring devices)?
	What supplies are available to them?

SRT	What is the mission of the SRT? Will the command post remain in its present location throughout the operation? Are lines of communication with the command post, forward observation post, and helicopters available throughout the mission? Are other law enforcement officers, firemen, military personnel, or civilians in the area aware of the situation, and are lines of communication established with them? Are detailed physical layouts of the target area available (maps, photos, building plans)? What are the best approach and withdrawal routes? Are there usable storm drains in the area? What is the best position from which to counter the actions of the offenders or otherwise accomplish the objectives of the mission? Is intelligence information available on ther hostile elements in the area? Are there snipers in the area? What additional support, including medical, is available?

Figure 6-23. Priority intelligence requirements.

6-15

WEAPONS AND EQUIPMENT		
En route to the objective	**At the objective**	**For control during the mission**
Maps	Standard weapons	Radio
Compass	Ammunition (standard and special)	Whistle
Ropes and grappling hooks	Special weapons	Spray paint
Storm drain openers	Binoculars	Flashlight
Wire or bolt cutters	Radio	Luminous tapes
Flashlight	Flashlights	Routine equipment, such as uniforms and weapons
Vehicle (if necessary)	Any equipment needed to gain entrance	Provisions for water and rations
	First aid equipment	
	Body armor	
	Handcuffs of flex cuffs	
	Protective masks	

Figure 6-24. Example of logistical requirements.

CHARACTERISTICS OF THE ASSAULT

Once the team reaches the inner perimeter, it should secure covered positions. Snipers must take a position that affords them an unobstructed view of the hostage site. The objective of the SRT is always to accomplish the mission with minimum gunfire. Indiscriminate firing of weapons is extremely dangerous, particularly in a highly congested metropolitan area. If deadly force cannot be avoided, the SRT should attempt to kill the offender with one shot or incapacitate him quickly.

Every hostage situation will present different specifics. No two assaults will be identical. However, in all assault situations, certain principles must be followed. Tactics for some types of situations are outlined in FMs 19-15 and 90-10.

Principles governing the assault include undetected approach, fast and furious entry, and offensive tactics, as well as accurate weapons firing. Shock, speed, and coordinated action provide the key to a successful assault. The assaulting elements must know where the offenders are, where the hostages are, and where the other members of the SRT are. Quick identification of these elements plus immediate reaction is essential. Simultaneous entry from two or more directions is preferable to single entry from one direction. This may be done by entry from the top down (blow a hole in the ceiling) and through a door or wall together with a simultaneous diversion from a point in the view of the offenders. Explosive entry must be made with great care to prevent serious injuries to the hostages. The use of flash and stun grenades, if available, can disorient the offenders momentarily, thereby reducing the likelihood of injury to the hostages. However, the assault team must react quickly to neutralize the offenders. Again, speed and decisiveness will increase the survival chances of the hostages.

6-16

GENERAL RULES FOR INFILTRATION	Demand perfect discipline within the team. A mission can be safely accomplished only by perfect teamwork with no individual heroics. No team members should ever leave the team except under orders. Never abandon wounded members of the SRT. Do not delay in the open. In built-up areas, use available cover and concealment. Select a new covered and concealed position before moving to another position. Protect the rear and flanks in all phases of the mission. Move rapidly from one position to another. Keep a safe distance between team members Do not mask covering fire. Keep talking to an absolute minimum. Communicate as much as possible by hand signals. Keep volume of radios as low as possible and use an earplug receiver, if available. All members of the SRT should have a means of communication. Maintain a positive communications link (countersnipers, in particular). Use visibility reducing conditions (darkness, fog, rain, wind) to conceal team movement to the objective. Know the complete order before departing, plus rallying and rendezvous points. If possible, go immediately to the roof or upper floors. Use a fire escape, inside stairs, or scale a wall. Stay away from front or target side of building. Use fire escapes cautiously. Persons on fire escapes are easily observed and therefore vulnerable. Avoid inside stairwells; they are ready-made for ambush. Ascend through any other reachable area, one man at a time, covered at all times. Establish an abort code word for SRTs and countersnipers prior to deployment.
SRT LEADER	Select the best routes and an alternate route for each member before beginning the infiltration. Select a route to the rear or on an unobservable side of the location of the offenders. Account for SRT members during the entire mission.
TEAM MEMBERS	Take care when moving across open ground, provide cover for other members who are moving, and ensure no one moves across open ground until the moving men are covered. Take advantage of natural cover and concealment (doorways, trees, storm drains, trash cans, vehicles, and shadows). Do not move along ridges and edges of roofs. Carry weapons in a ready position. Take advantage of diverting noises while moving at night (wind, vehicles, airplanes). Do not move on roads or streets unless absolutely necessary. In an alley, move along the side with the darkest shadows. Watch out for trash cans and trash. Avoid lateral movement across the forward position of the terrorists' position.

Figure 6-25. Rules for infiltration.

PLANNING AN ASSAULT

Formulate a plan for approaching the objective and then conduct exhaustive realistic rehearsals that include implementation of contingency plans. Rehearsals should include, if possible, both blank fire and live fire execution during daylight and reduced visibility. During rehearsals, assign individual team member responsibilities.

6-17

Initiate the planning and rehearsal phase well in advance of the decision to use force. Base all plans and actions on the principle of surprise by achieving clandestine closure with the target and by using unexpected actions or techniques.

Orient the SRT planning and rehearsals on the *one team – one mission* concept.

Avoid a chain of events concept that requires one team to accomplish more than one mission.

Rehearse the assault exhaustively to ensure that it is coordinated and accurately timed. Once the assault team is committed, the SRT commander should be in total control and absolutely free from outside interference.

The following general principles, though not all inclusive, are good guidelines for the SRT to approach the terrorists and gain entry to a building or to establish observation positions:

- Select a point or, preferably, points for entry before movement.
- Determine a safe route to the point(s) of entry.
- Deploy SRT members to cover the approach team.

Consider creating a diversion to distract the offenders' attention from the approach team, such as loud talking on the public address system, placing another telephone call, or flying a helicopter or other aircraft over the side of the location opposite the approach route.

Consider employing tear gas. The offenders' self-preservation instincts and the confusion that results during a gas attack may sufficiently distract them long enough to prevent them from carrying out any threats. The use of gas, however, often has as many negative effects as positive ones. If smoke is used simultaneously with tear gas, it will provide excellent concealment for SRT members maneuvering into position to kill or incapacitate the terrorists. However, gas or smoke should be used only in support of an SRT assault. The success of this type of operation depends on speed, cool thinking, teamwork, and training.

If gas is to be used, keep in mind the potential fire hazards, the effects of wind, the area where the gas is used, and unprotected personnel and hostages in the operational area. Diverting the offenders' attention from the point of gas entry will help to assure the proper detonation of the gas before the offenders can throw or kick it away.

Use smoke as concealment of the approach team if no other concealment is available. If smoke is used, the SRT members should wear gas masks.

Select the least vulnerable points in the structure for entry. When possible, enter at the top of the building and move downward through the building. Avoid entrance through windows and doors, if possible.

Ladders, drainpipes, vines, helicopters, or the roofs and windows of adjoining buildings may be used to reach the top floor or roof of another building. Or, one member may climb onto a buddy's shoulders to pull himself up. Scaling ropes are also useful. By attaching a grappling hook to the end of a scaling rope, a team member can scale a wall, swing from one building to another, or gain entrance through an upstairs window.

Rappelling from buildings and helicopters is another method of gaining entrance to the upper levels of a hostage site.

Divide the SRT into a support team and an assault team when assaulting a building. Each team normally consists of two or three persons. The support team provides continuous firepower protection for the assault team. Prior to the assault, every room in the building may need to be cleared. This is the mission of the assault team. During the clearing of a building, if the hallway must be used, avoid presenting a large target to the enemy, and get out of the hallway as soon as possible. Also, when moving within a building, avoid being silhouetted in doorways and windows.

If forced entry into a building or a barricaded area is decided upon, the entry team must take appropriate equipment, such as a pry bar, sledge hammer, battering ram, gas gun, projectiles, explosive breaching device, hand grenades, and smoke grenades. The entry team should wear body armor, if available, and gas masks if gas or smoke is to be deployed.

To mechanically breach a locked wooden door, a jimmy tool or crowbar can be used to pry the door open. If the door is steel, the lock cylinder can be twisted off with vicelike tools and the locking device punched out or the locks can be picked.

An even more sophisticated and expeditious method of forced entry is to use explosive devices that blow the lock out of the door. Using a shoulder weapon to shoot the lock out is a last resort. The demolition team can also place demolitions on the door hinges so that the door falls forward, keeping the entrance clear.

Note that mechanical breaching of the barricaded area may be too slow to achieve surprise. If this method is required, a backup system should be on hand in case the primary method fails. Whatever method is used, simultaneous entry through multiple breaching points is essential.

If entry is to be made through a window, team members should try to find an open or unlocked window. If a secured window is the point of entry, completely break it out to avoid injury from jagged glass. Make every effort to be as silent as possible, or if the situation warrants it, as fast as possible. Use a periscope device, if available, to observe the interior for offenders or any booby traps or channeling obstructions.

Team members should make maximum use of the architectural features of the building for movement, such as heating and ventilating shafts, false ceilings, and elevator shafts.

Once the openings are made, the SRT members must effect simultaneous multiple entries into the target area. The offenders must be immediately threatened from several points simultaneously. Their first reactions will probably be self-preservation, which will help prevent the immediate killing of hostages. The assault team must ensure it presents a threat to every offender within 8 to 15 seconds of their detection of the attack. The most useful technique is to use nonlethal weapons or devices that create a shock effect, for example, demolitions and concussion grenades.

To eliminate exposure time, team members should move quickly through any openings. The team members should enter simultaneously with each member having a predetermined sector of fire. The first assault team members should seek cover in a predetermined location in the room. If cover is unavailable, they should squat in a combat firing position, minimizing their vulnerability as a target. The next assault team members should move simultaneously into the opposite sides of the target area. By taking up positions on opposite sides of the target area, reaction team members can provide direct fire on any offenders, if necessary. Team members must close with and eliminate the threat within the shortest possible time.

6-19

If the offenders' location is unknown, the team members entering the building should point their weapons toward the portion of the room not in their sight, as it is likely that the offenders will position themselves here they cannot be hit by direct fire through the doorway.

Since the immediate identification of the offenders is essential, maximum use should be made of the information from the priority intelligence requirements. To further differentiate the offenders from the hostages, remember—

- The offenders most likely will have weapons while the hostages will be unarmed.
- The offenders probably will be more active while the hostages will probably be inactive and sitting or lying down.

The SRT must secure any hostages within 30 to 45 seconds. If terrorists are highly trained, the team may have only 5 to 7 seconds. Hostages must be positively identified and immediately removed from the danger area. The SRT must understand that due to the violence of execution, the hostages may initially fear the SRT more than the offenders. Additionally, SRT members must be wary of offenders who pose as hostages.

After entry is gained and the search of the immediate area is complete, the SRT should search and clear the remainder of the structure using the same method. The team must make every effort to have the whole target area entered at the same time, avoiding any piecemeal actions. The remainder of the building should be cleared as rapidly as possible.

To ensure the whole building is cleared, each room must be entered, neutralized, and secured before the entry team goes into any adjoining rooms. As each room is secured, the entry team positions a person in that room to assure no offenders move in after the team has left. These additional personnel should be available for deployment from the command post as needed.

The same security precautions should be made with anyone found within the objective area. Use a positive method to signify which personnel have been searched for concealed weapons. One technique is to spray paint cleared personnel.

Immediately following the assault, the SRT should conduct a thorough, detailed briefing to—

- Emphasize the strong and weak points of the operation.
- Recommend improvements.
- Determine the effect of the tactics employed.
- Determine the operational effects of the equipment.
- Discuss other pertinent matters.

ASSAULT TRAINING

Training should be an ongoing process. It should be structured to ensure that each team member is physically and mentally prepared to cope with high-risk situations. Such training is necessary to minimize injury or death of fellow team members or other personnel. To achieve this goal, each team member must be aware of what other team members will do under certain circumstances.

Training Diversification

The SRT requires diversified training. In addition to rigorous physical training and extensive firearms training with a variety of weapons, SRT training should include the history

6-20

of guerrilla warfare, scouting and patrolling techniques, night operations, camouflage and concealment, counterambush techniques, rappelling, chemical agent use, and first aid.

Subjects that are taught in a classroom setting should be tested under simulated high-risk situations in the field. Psychological training is extremely valuable to SRT members. Such training helps team members to understand how a terrorist, a hostage taker, or an offender might react, what motivates him, and how best to deal with him without employing deadly force. Also, such training helps team members to better understand how they might react individually under pressure.

Training courses in SRT tactics should be conducted in an initial 6- to 8-week intensified training period to develop skills in tactics and theory. Following this initial training period, training may be conducted an average of five days a month to refine these basic skills. The Federal Bureau of Investigation (FBI) conducts similar SRT training.

Marksmanship

The environment in which an SRT must operate places a premium on accurate, semiautomatic fire. The typical target will present itself for a fleeting moment, possibly at a long range. For this reason, basic rifle marksmanship must be a high priority for SRT members. In addition to rifle marksmanship, each SRT member should be trained on all weapons used by the SRT, including sniper weapons, pistols, and submachine guns. Certain members should train more extensively on specific weapons to become expert in their use. All members should be trained to fire both the pistol and the submachine gun instinctively and accurately with aimed fire.

In addition, to take advantage of maximum cover in urban areas and facilitate firing from corners, SRT members must learn to fire from both shoulders. For example, right-handed firers must learn to fire from the left shoulder and vice versa. Marksmanship training should include firing from both shoulders so that team members can become comfortable with it.

The searching and clearing of buildings and the techniques necessary in a final assault involve many quick, short-range engagements. Quick-fire techniques, as described in FM 23-9, should also be emphasized.

Sniper training is essential for the SRT. This training is covered in TC 23-14.

Physical Training

Members of the SRT must be in good physical shape, particularly in upper body strength and agility. Team physical training should emphasize the following subjects, which appear in FM 21-20:

- Rifle and Log Drills, Chapter 11.
- Strength Circuits, Chapter 15.
- Obstacle Courses, Chapter 16.

Individual Movement

A basic consideration for the success of the team is the ability of each team member to move efficiently from one place to another. Individual movement may mean the difference between success and failure of the mission. It may make the difference between life and death. Each member must be prepared for the mission and must thoroughly understand individual movement techniques.

6-21

Each member should prepare himself and his equipment by camouflaging his face, hands, neck, backs of ears, and his equipment. He should tape loose metal equipment to prevent it from rattling, and he should wear soft, well-fitted clothes. Trousers may be tied at the thighs and at the ankles if their looseness would otherwise cause noise. Clothing must be loose enough, however, to ensure proper circulation. Wear a baseball cap; a helmet is clumsy, interferes with hearing, and gets in the way of precision shooting. Carry exactly what is needed; do not carry extra equipment. Wear black leather gloves, preferably unlined, for protection against broken glass, jagged metal splinters, and thorns. Each team member should carry his own equipment and extra ammunition. Each member should know his assignment, the objective of the mission, details of preparation, what his job is, what the jobs of the other team members are, what the rallying and rendezvous points are, the primary and alternate routes, and the call signals and passwords. The *buddy system* should be used to check each other's gear, including the team leader's.

Individual movement should be by bounds, halting, listening, observing, and then by moving again.

The team leader should select the next spot before leaving the concealment of an occupied position. He should carefully observe the area for hostile activity, then select the best available covered and concealed route to the new location. Take advantage of darkness, fog, smoke, haze, and noise to assist in concealing movement.

Team members should remain alert upon reaching a new position, and they should listen carefully. They should also observe briefly to see if their movement alarms animals. The flight of birds, movement of small animals, or barking of dogs may alert the offenders.

Movement should be made across roads and alleys where there is the most cover and concealment. Team members should look for culverts, low spots, curves, parked autos, trash cans, weeds, bushes, and trees.

Furrows or natural indentations in the ground should be followed when crossing vacant lots or fields. Even the slightest depression can provide cover and concealment from ground observers.

Team members should be constantly aware of the three-dimensional aspect to urban hostilities; they are vulnerable from all sides, on the ground, and from windows, trees, and rooftops.

Steep slopes and areas with gravel or loose stones should be avoided.

If possible, freshly watered lawns should be avoided while reconnoitering; tracks permit the enemy to know the team was there.

Where possible, avoid clear areas to eliminate being silhouetted against the skyline.

Use the *rush* to go from one position to another when speed is essential. The correct technique is as follows:

- Start from the prone position.
- Slowly raise your head and select a new position.
- Slowly lower your head, draw your arms into your body, keeping your elbows down and pull your right leg forward.
- Raise your body with one movement by straightening your arms.

6-22

144

- Spring to your feet, stepping off with your left foot.
- Run to the new position using the shortest route. Do not try to move to a new position that is too far away from the original position.
- Plant both feet just before hitting the ground.
- Drop to your knees, at the same time sliding your right hand to the heel of the rifle butt.
- Fall forward, breaking the fall with the butt of the rifle.
- Shift your body weight to the left side. With your right hand, place the butt of the rifle in the hollow of your right shoulder.
- Lie as flat as possible. If there is reason to believe your movement was observed by the offenders, move to the right or left (if cover and concealment exists).

Use the crawl when it is necessary to move your body close to the ground to avoid being seen. Two ways to crawl are the low crawl and the high crawl. Use the best way suited to the conditions of visibility, cover and concealment, and speed.

Use the low crawl when cover and concealment are scarce, when visibility permits good observation by the offenders, and when speed is not essential. Keep your body as flat as possible against the ground and, if using a rifle, grasp the rifle sling at the upper sling swivel. Let the balance of the rifle rest on the forearm and let the butt of the rifle drag on the ground. To start forward, push with your arms forward, pull your right leg forward (flat on the ground), pull with your arms and push with your right leg. The pushing leg should be changed frequently to avoid fatigue. Make every effort not to raise or lift any part of the body (particularly the buttocks). Another technique for transporting the rifle is to have the rifle slung over your back and one shoulder. If a weapon other than a rifle is used, modify the above technique to accommodate the situation.

Use the high crawl when cover and concealment are available, when poor visibility reduces the offenders' observation, and when more speed is required. Keep your body free of the ground and rest your weight on your forearms and lower legs. Cradle the rifle in your arms, keeping its muzzle off the ground (if right-handed), butt of rifle to the right, and muzzle to the left. Make the move forward by alternately advancing your right elbow and left knee, left elbow and right knee.

Use the following walking techniques when extreme quiet is necessary:

- Make footing sure and solid by keeping your weight on one foot as each step is taken.
- Raise the other leg high to clear brush or grass or rubble.
- Gently let your front foot down, toe first, with the weight on the rear foot.
- Feel with your toe to pick a good spot.
- Lower your heel after finding a solid place.
- Shift your weight and balance to your forward foot.
- Take short steps to avoid losing balance.
- If you are operating in the dark, hold the weapon in one hand and extend the other forward, feeling for any obstructions. Storm drains are a good example of a place where it is pitch black and flashlights cannot be used for fear of warning the offenders.

6-23

Assume the prone position by the following methods:

- Crouch slowly to assume the prone position.
- Hold the weapon under your arm and feel for a clear spot with your weight on your free hand and opposite knee.
- Raise your free leg up and back, and lower it to the ground, feeling with your toe for a clear spot.
- Roll gently into the prone position.

Use the following techniques when crawling on your hands and knees:

- Do not use the low and high crawl when very near the hostage site because of the shuffling noises.
- Crawl on your hands and knees and lay the weapon on the ground by your side.
- Make a clear spot for your knee by feeling the ground with your right hand. Keep one hand on the spot and bring your right knee forward until it meets your hand.
- Repeat the same procedure with your left hand and knee.
- Move the weapon by feeling the ground for a clear place and lifting the weapon into the new position.
- Crawl very slowly, staying as low as possible. Remain absolutely silent.

Use the following methods for searching terrain:

- Take care to prevent telltale reflection from the lenses when binoculars are used.
- Search by shifting your eyes with short, jerky movements.
- Limit the scan to a particular area if something is observed.
- View an object against the sky from a position close to the ground as it can be more clearly seen from that position than from any other position.
- Do not look over the top of objects unless absolutely necessary. Be sure that the background breaks up the silhouette. The best technique is to look around the sides of objects, such as fences and buildings, from the lowest level possible.
- Look slightly to the side of an object you wish to see at night, not directly at it.
- Do not look into bright lights (sun, fire, headlights) at any time. If the mission begins in sunlight and carries into dark hours, sunglasses can help transfer to night vision. Remember that sunglasses can also reflect sunlight.
- Avoid looking over or around skylines or walls. Use natural openings such as drain holes, broken walls, and curtained windows. Take advantage of natural cover and broken outlines or background.
- Use the correct technique for looking around a corner. The correct technique is to lie flat on the ground, ensuring that your weapon is not extended beyond the corner of the building and that your head is exposed at ground level only enough to permit observation around the corner. A common error is for a right-handed firer to fire from his right shoulder around the left corner of a building. The firer should fire from his left shoulder to present the smallest possible target.

Additional considerations in individual movement are as follow:

- Do not smoke while on a mission unless specifically announced otherwise.
- Do not conduct individual movement against a sniper any closer than necessary, but get as close as possible.
- During movement, do not select a position from which effective fire cannot be immediately delivered.
- Check for booby traps, then cut any wire near the picket to prevent loose ends from flying back. Start on the lower strands first.
- Crawl up to the trench and place your body parallel to it (if right-handed, left side to the trench), to observe the inside of the trench. Lower your left leg into the trench, then slide your body over the side, hitting feet first with your head down. Be alert for fire and move down the trench a few yards before getting out. When leaving a trench, quickly observe over the edge of the trench.
- Be aware of background and try to blend in with it. Avoid contrast if possible. Members must take care not to silhouette themselves and should keep low at all times.
- Step high and come down flat-footed for silent walk.
- Additional information on individual movement is available in FM 21-75.

When using a grappling hook and rope, take care to select a hook that is sturdy, portable, easily thrown, and equipped with prongs that are likely to hold inside a window. The rope should be 5/8 inch to 1 inch diameter and long enough to reach the targeted window. Knots tied in the rope at 1-foot intervals will facilitate climbing.

When throwing the grappling hook, stand as close to the building as possible. The closer the thrower stands, the less he is exposed to fire from different directions. At close range, the hook has to be thrown less horizontal distance.

When throwing a grappling hook, allow the rope to play out freely. The grappling hook thrower should have enough rope to reach his target. In his throwing hand, he should have the hook and a few coils of rope. The remainder of the rope, in loose coils, should be in his other hand. The throw itself should be a gentle, even, upward lob of the hook with the thrower's other hand releasing the rope as it plays out.

After throwing the grappling hook, ensure that the grappling hook has a solid hold before beginning to climb. Once the grappling hook is attached, pull on the rope to obtain a good hold. When using a window, pull the hook to one corner to ensure your chance of a good "bite" and to reduce exposure to lower windows during the climb. Team members should avoid silhouetting themselves in windows of uncleared rooms when scaling walls, and they should enter an upper window with a low silhouette. Entry can be head first; however, the preferred method is to hook your leg over the window sill and enter sideways, straddling the ledge.

Additional information on climbing techniques can be found in TC 90-6-1.

In the seat-hip rappel, the main friction is taken up by a snaplink, which is inserted in a sling rope seat fastened to the body. The seat-hip rappel is a fast method of getting down a wall and is also used in rappelling from helicopters.

To attach the seat and hook snaplink, place the sling rope across your back so that its center is on the hip opposite the hand that is used for braking. (If the rappeller is right-handed, the right hand is the brake hand or vice versa.)

6-25

147

Tie an overhand knot in front of your body. Bring the rope between your legs (front to rear), around your legs, and under your waist loop.

Tie the ends with a square knot and two half-hitches on the side opposite your brake hand.

Place the snaplink through the single rope around your waist and through the two ropes, forming the overhand knot.

Insert the snaplink with the gate down and opening toward your body.

Rotate the snaplink one-half turn so that the gate is up and opens away from your body. To rappel, stand to one side of the ropes. (When braking with your right hand, stand on the left side; when braking with your left hand, stand on the right side.) Snap the ropes into the snaplink.

Take up some slack in the ropes between the snaplink and the anchor point and bring the rope underneath, around, and over the snaplink, snapping into it again. (This results in a turn of the ropes around the solid shaft of the snaplink so they do not cross themselves when under tension. When using a single rope, make two turns.)

Back carefully over the edge of the obstacle, face the anchor point, and lean well out, almost at a right angle to the surface (the *L position*).

Go down using your upper hand as a guide and slightly above your hip. Brake by closing your hand and pressing the rope against your body.

Continue to *walk* down, looking at the ground over your brake hand. As confidence is gained, you can go faster by pushing off the surface and going down in *bounds* with your brake hand extended toward the ground.

Tuck the fatigue jacket into your trousers before rappelling, as loose clothing or equipment around your waist is apt to be pulled into the snaplink, locking the rappel.

Ensure the rope reaches the bottom or a place from which further rappels can be made.

Test the anchor point carefully and inspect to see that the rope will run around it when it is pulled from one end from below (in order to recover the rope). Ensure that the area is clear of loose rocks that may be pulled off. Look for bee and wasp nests.

Give the signal *off rappel* when the bottom is reached and straighten the ropes. (When silence is required, use a prearranged signal of pulling on the rope.)

Recover the rope once the last man is down. Pull it smoothly to prevent the rising end from whipping around the rope. Stand clear of the falling rope.

Additional methods, techniques, and information on rappelling are available in TC 90-6-1.

CHAPTER 7

FIELD–EXPEDIENT PRINTING TECHNIQUES

Special Forces, and the foreign indigenous groups with which they work, often require printed material to support training, information dissemination, and psychological warfare requirements within the area of operations. In certain situations, the production of printed material is of primary importance to the success of the mission. Because the occupying power usually imposes strict controls on printing material and equipment, the production of printed material often requires that field–expedient printing techniques be used. The field–expedient printing techniques described in this chapter can be used to meet operational requirements.

MAKING AND USING A SILK SCREEN

The field printer must either carry or make the tools needed for printing matter any time, anywhere.

Tools For the Job

A silk screen, a stencil, ink, a stylus, paper, and a squeegee are necessary for printing in the field. The field printer can carry them along whenever he expects to do printing in the field; but a good workable substitute for all of these tools can be found in any forest, swamp, or desert.

Silk screen. A silk screen (Figure 7–1) consists of a frame over which a piece of fabric is stretched. This frame is attached to a base to provide a flat working space. The cover is not necessary for printing but simply makes the silk screen easy to carry from one place to another.

Stencil. A stencil is a device that allows the ink to pass through the screen and onto the paper where it is needed and blocks out the ink where it is not needed.

Ink. The ink used in silk screen printing should be thick and have an oil base. Many kinds of ink can be used for printing in the field.

Stylus. A stylus is a device used to etch the stencil. A pointed piece of wood or metal can be used for this purpose.

Paper. Paper, or a good substitute, is an essential item for printing in the field. Many good substitutes for paper have been found, but it is best to have a good supply of paper whenever possible. Often paper that has been used can be reused by the printer for a new mission.

Squeegee or ink roller. A squeegee, or ink roller, is a tool used to spread ink evenly and to force the ink through the stencil and onto the paper.

Silk Screen Printing Base and Cover Construction

The field printer can construct a silk screen printing press by following the instructions in Figures 7–2 through 7–5. Remember that the silk screen and all of the other items

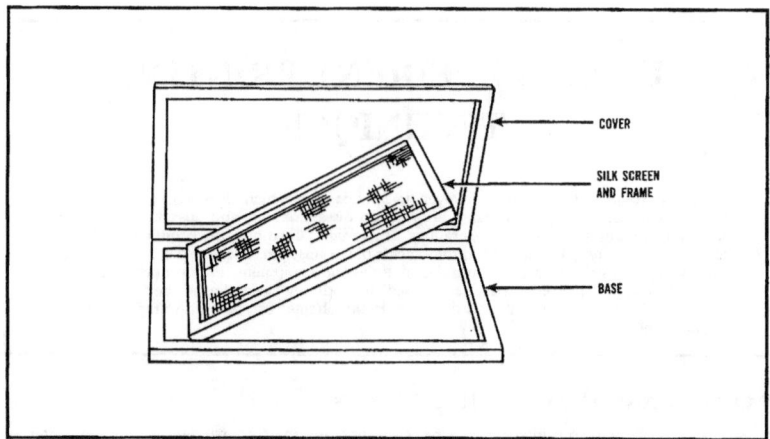

Figure 7-1. Silk screen with carrying case.

mentioned can be made by using materials found in the field. A good serviceable silk screen can be made by using wooden pegs instead of nails, a rock instead of a hammer, a knife instead of a saw, and bamboo instead of pieces of wood for the frame. Nails must be very thin so they will not split the wood. It is best to use *soft* wood in making the frame.

Many kinds of fabric can be used to make the screen. However, silk fabric gives the best results. It is strong and can be cleaned and used many times. Parachute nylon or a cotton handkerchief can serve in an emergency. Even an undershirt can be used; however, remember that only finely woven fabrics will allow fine lines to be printed.

The Ink to Be Used

Many different kinds of ink can be used for printing with the silk screen. Ink with an oil base, such as mimeograph ink, is best. Paint with an oil base is the best substitute, or printer's ink can also be used. Ink that is used for silk screen printing should be thick; oil base paints are almost the right thickness. By experimenting with many kinds of ink, the printer will learn what to look for in a good printing ink. In an emergency, the field printer can crush berries or any stain producing material and make an ink substitute.

How to Use the Stencil and Silk Screen

Place the words, picture, or symbols on the stencil. If you are using the standard printing stencil, scratch the words onto the stencil with the pointed stylus. If you are using the cut-out stencil, remove the parts with a knife or sharp object.

Lift the silk screen frame up from the base as in Figure 7-1. Place the stencil on the bottom of the screen. Tacks, tape, or glue can be used to hold the stencil in place.

Place a piece of paper on the base under the stencil. This piece of paper serves to protect the base from ink while preparing to print.

Lower the silk screen onto the base. Place enough ink on the silk to cover the screen. Use the squeegee to spread the ink evenly and to force the ink through the openings in the

Tools for making a silk screen:

- A hammer or heavy object for driving tacks and small nails
- A knife for cutting the fabric and canvas hinge
- A saw or hatchet for cutting the wood

Materials for constructing a 22 1/4" X 16 1/2" frame:

- Four pieces of wood, 1 1/4 X 3/4 X 15 1/4 inches
- Four pieces of wood, 1 1/4 X 3/4 X 21 inches
- Sixteen 1-inch nails
- Two 1 1/4-inch nails

Figure 7-2. Materials and measurements for constructing a silk screen.

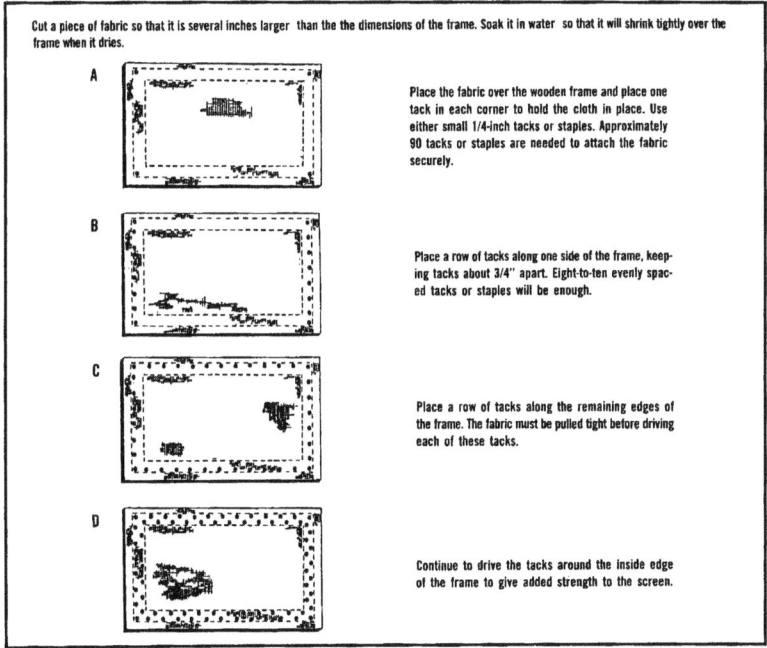

Cut a piece of fabric so that it is several inches larger than the the dimensions of the frame. Soak it in water so that it will shrink tightly over the frame when it dries.

A Place the fabric over the wooden frame and place one tack in each corner to hold the cloth in place. Use either small 1/4-inch tacks or staples. Approximately 90 tacks or staples are needed to attach the fabric securely.

B Place a row of tacks along one side of the frame, keeping tacks about 3/4" apart. Eight-to-ten evenly spaced tacks or staples will be enough.

C Place a row of tacks along the remaining edges of the frame. The fabric must be pulled tight before driving each of these tacks.

D Continue to drive the tacks around the inside edge of the frame to give added strength to the screen.

Figure 7-3. Tacking cloth to underside of silk screen frame.

7-3

151

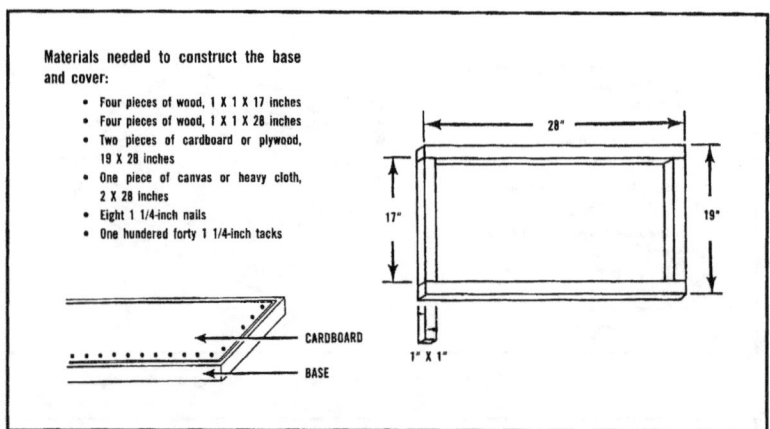

Materials needed to construct the base and cover:

- Four pieces of wood, 1 X 1 X 17 inches
- Four pieces of wood, 1 X 1 X 28 inches
- Two pieces of cardboard or plywood, 19 X 28 inches
- One piece of canvas or heavy cloth, 2 X 28 inches
- Eight 1 1/4-inch nails
- One hundered forty 1 1/4-inch tacks

Figure 7-4. Materials and measurements for constructing base and cover.

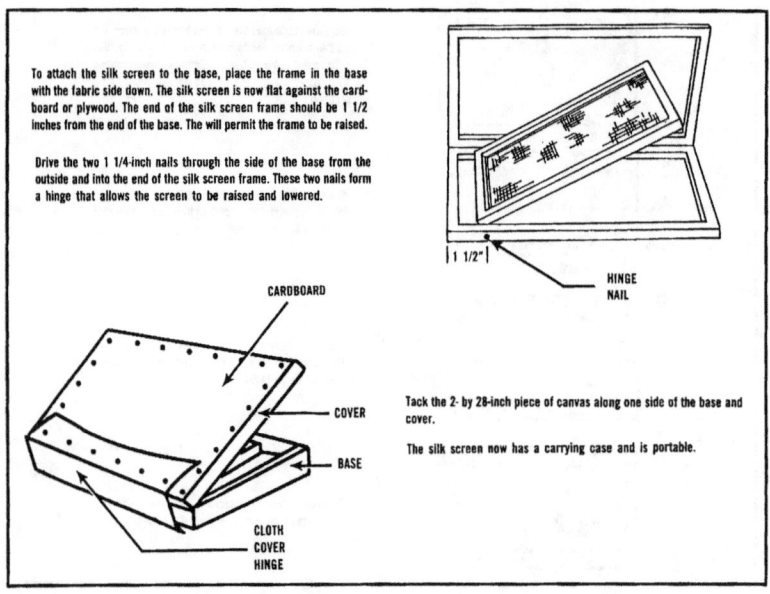

To attach the silk screen to the base, place the frame in the base with the fabric side down. The silk screen is now flat against the cardboard or plywood. The end of the silk screen frame should be 1 1/2 inches from the end of the base. The will permit the frame to be raised.

Drive the two 1 1/4-inch nails through the side of the base from the outside and into the end of the silk screen frame. These two nails form a hinge that allows the screen to be raised and lowered.

Tack the 2- by 28-inch piece of canvas along one side of the base and cover.

The silk screen now has a carrying case and is portable.

Figure 7-5. Making hinges for the silk screen kit.

stencil. The squeegee must have a straight edge. Another tool that will do the same job is a roller. A roller made of hard rubber is best for spreading the ink on the silk screen. A stiff brush is another tool that can be used.

To print, place the piece of paper to be printed on the base and lower the silk screen on top of the paper. Slide the squeegee firmly over the silk, forcing the ink through the stencil. Lift the screen, remove the paper, and allow the paper to dry. If the printing is not dark enough, add more ink to the screen.

The field printer can print words, pictures, or symbols. Note that the first step is to ensure that all tools are clean and in good working order and that there is enough paper to finish the job.

When the printing job is finished, remove the stencil and clean the screen and all other tools. Also, be sure to clean the squeegee.

MAKING AND USING A ROCKER-TYPE MIMEOGRAPH MACHINE

Cover any smooth, curved surface with a heavy (thick), porous fabric. Saturate the fabric with mimeograph ink. Cover the ink pad with the desired stencil and apply to appropriate paper with a rocker-type movement of the apparatus.

By using many ordinary items, an inking base for a rocker-type mimeograph machine can be made with crude tools or in some cases, the item may be used as it is. Any smooth surface, such as a tin can or glass bottle can be used as a base. A larger frame can be made from a wooden block, using a chopping axe and a penknife. The wooden block can be hollowed out to carry ink, styli, and an extra supply of stencil paper. Size can be increased by fastening a piece of sheet metal to the block.

An *inking pad* can be made by using thick, porous materials such as a coat, a blanket, felt, or burlap. A pad also can be made of many layers of thin fabric. Wrap the pad around the smooth, curved surface of the base. The pad can be held in position with tape, string, thumb tacks, or glue. Saturate the pad with mimeograph ink.

This *ink* can be a composite of almost any grease and carbon scraped from a fireplace or grating. Color can be achieved by mixing pigments of color to the grease instead of carbon. Mimeograph ink, commercial grade, is a universal item and is available in many countries. Shoe polish, thinned with kerosene or other solvent, is generally available and usable.

Stencils can be made from thin, tough tissue or thin airmail paper by applying a coat of wax (paraffin) to one side. This wax can be rubbed on, then gently warmed to ensure uniformity of thickness and penetration of the paper. Only partial penetration is desirable, not saturation.

For a *stylus*, use a ballpoint pen, a slender stick of hardwood, or even a heavy piece of wire with the ends rounded and smoothed enough to etch the wax without tearing the paper. Use the stylus to inscribe the desired message or to sketch on the wax coating of the stencil. Then apply the stencil to the ink pad with the wax next to the ink. Some of the ink will penetrate through the lines made by the stylus, thus *inking* the stencil. The undisturbed wax prevents the ink from penetrating the paper in unwanted places. To print, lay the stencil on a sheet of paper with the inked surface next to the paper. Rub the back of the stencil to transfer the ink to the blank paper.

7-5

If no *mimeograph paper* is available, substitute paper should be of quality equal to newsprint, but almost any paper will suffice.

MAKING AND USING A GELATIN PRINTING DEVICE

This reproduction method is more commonly known as the hectograph technique, a commercial technique used worldwide. All necessary materials are commerically known by the name *hectograph* and are available in several variations from gelatin plates to prepared plates which are fiberbacked, wraparound models for machine use (Ditto). The Ditto machines are similar in appearance to mimeograph machines. Emergency or field conditions will probably dictate the use of a simple gelatin plate.

Gelatin, which is the base for this technique, can be purchased as a hectograph product, made from gelatin powder produced by food concerns (such as Knox), or made by boiling the bones and skin of animals. (Pulverizing the bones will speed the boiling down process.) Add enough gelatin powder to make a semisolid plate. Pour the warm liquid gelatin into a shallow, wide container or on a tabletop and allow it to cool and set. When properly prepared, it becomes a glass–smooth plate that feels like sponge rubber to the touch. This plate will be soft enough to absorb the ink but firm enough not to bleed the ink on the master copy. The addition of a little animal glue will toughen the plate and a little glycerine will keep it from drying out too quickly. The effects of these additions are in direct proportion to the quantity used; both are desirable, but not absolutely necessary. Both should be added and well mixed during the liquid stage of the gelatin.

Make the master copy using a good grade of smooth, tough, hard–finish paper. Using hectograph or Ditto carbon paper, ribbon, ink, or pencil (all are commercially available), type or write the material to be reproduced. Trial and error testing will unveil numerous ink pencils (indelible), writing inks, and stamp pad inks that will reproduce. Do not blot after applying the ink to the master copy. If pencil is used, be sure that the copy is strong and uniform.

When the gelatin plate is set and ready for work, sponge the plate thoroughly with cold water and allow it to set for an additional minute or two. Using a sponge, remove all excess moisture and apply the master copy, face down, on the gelatin plate. Carefully smooth the copy to ensure complete and uniform contact with the prepared plate. Do not remove for at least 2 minutes. Lift one corner of the master for a gripping point and smoothly and carefully lift the master copy from the gelatin plate. The gelatin plate now bears a negative copy of the desired material and is ready to reproduce the copy.

Begin reproduction immediately after the master copy has been removed from the gelatin plate. Place a blank sheet of smooth surface paper on the gelatin plate. Using one hand (or a rubber roller, if available), smooth it into total contact. Then lift the sheet from the gelatin surface. This is done rapidly to obtain as many copies as possible from one inking of the plate. One good inking of the plate may produce from 100 to 200 copies using this method, while a commercial Ditto machine may produce as many as 700 copies. To speed this process, leave one small corner of the sheet of reproduction paper free for gripping. This can be done by sticking a small piece of paper to the place on the gelatin plate where a corner of the reproduction paper would fall. This piece of paper acts as a guide and a buffer to keep that one corner of the reproduction paper from sticking. When removing the reproduction paper, lift the sheet by the loose corner; do not attempt to roll it away. The rolling action will cause the reproduction paper to curl as it dries.

After completing the reproduction job, sponge the gelatin plate thoroughly with cold water and allow it to set for 48 hours or until the ink has been assimilated by the gelatin. The

7-6

plate is now ready to be used on a new and different job. The only way to shorten the waiting time between jobs is to dissolve the gelatin plate in hot water, boil off the excess water until the liquid is thickened to the desired consistency, and pour a new gelatin plate. Of course, two or more gelatin plates may be prepared to increase production capabilities.

CHAPTER 8

PHYSICAL SECURITY AND VULNERABILITY ASSESSMENT

Antiterrorism is one of the major roles of US Army SOF in terrorism counteraction. Antiterrorism refers to the defensive measures used to reduce vulnerability to terrorist attacks. These measures include physical security and vulnerability assessments, which help safeguard personnel from terrorist acts. These measures also prevent unauthorized access to equipment, facilities, materiel, and documents. Physical security programs implemented according to regulations deter or reduce the chances of successful terrorist attacks. A successful physical security program includes planning, coordinating, executing, reviewing, and evaluating courses of action that improve physical security of quarters, offices, and installation facilities. Physical security includes human resources and mechanical systems that complement other installation programs such as crime prevention and operations security (OPSEC).

PHYSICAL SECURITY MEASURES

Physical security of an installation protects government information and resources. The more an area is strengthened, the more difficult it is for terrorists to penetrate. Intrusion detection systems, proper use of lighting and fences, authority to close access to an installation, security sensitive storage locations, and well-trained security personnel are all measures that enhance physical security. Additionally, a good crime prevention program provides the threat analyst with information about crime statistics at the installation level. Crime prevention programs identify, control, reduce, eliminate, or neutralize conditions conducive to crime. Generally, because a terrorist uses criminal methods, proven crime prevention techniques can be successful against a terrorist.

Objectives

The objectives of physical security are to—

- Use Department of the Army (DA) and major Army command (MACOM) threat statements to develop a local threat statement.
- Develop physical security plans, surveys, and inspections.
- Monitor work orders.
- Involve installation staff (threat committee).
- Establish and implement priorities for action.

Security Planning Measures

Thorough preparation is needed for the successful completion of any operation. This need is accentuated in a special threat situation because of the gravity of the consequences. Figure 8–1 is a security planning checklist that illustrates measures to take as precautions against the threat of terrorism. These measures must be continually reevaluated and updated.

IDENTIFY AND ANALYZE POTENTIAL PROBLEM AREAS	☐ Installation vulnerability assessment ☐ OPSEC survey ☐ Physical security survey ☐ Personal security survey
DEVELOP CONTINGENCY PLANS THAT COUNTER THREATS TO AN INSTALLATION	☐ Bomb ☐ Arson ☐ Armed assault, ambush, and assassination ☐ Hostage taking or subject barricading
DEVELOP RESOURCES (IDENTIFY, OBTAIN/ ESTABLISH, TRAIN)	☐ Equipment ☐ Special reaction teams (SRTs) ☐ Negotiators
COORDINATE INTERNALLY	☐ G1 through G4 ☐ Special staff
COORDINATE EXTERNALLY	☐ FBI ☐ Local law enforcement
DEVELOP LIAISON AND WORKING COORDINATION WITH KEY PLAYERS	☐ Fire department ☐ Hospital ☐ Legal ☐ Explosive ordnance detachment ☐ Aviation ☐ Public affairs officer ☐ Criminal Investigation Command ☐ Military intelligence ☐ Communications ☐ Facilities engineers ☐ Transportation motor pool
ESTABLISH APPLICABLE SOPs	☐ Notification ☐ Report ☐ Support ☐ Communication ☐ Transportation
TRAIN	☐ Special reaction teams (SRTs) ☐ Threat management forces (TMFs) ☐ Crisis management teams (CMTs) ☐ Negotiators ☐ Personal security guards (high-risk personnel)
EXERCISE, EVALUATE, AND ADJUST	☐ Communication exercises ☐ Command post exercises ☐ War games ☐ Field training exercises (FTXs) ☐ Alerts

Figure 8-1. Security planning checklist.

8-2

Special Planning Considerations

Figure 8-2 lists the areas that must be addressed when developing a contingency plan. As a minimum, the contingency plan should allow for a 7-day operation.

VULNERABILITY ASSESSMENT

While there is no absolute protection against terrorism, reasonable safeguards, commensurate with the identified threat, can be taken to reduce the likelihood of attack. Vulnerability of a particular target can be determined by identifying potential or actual terrorist activity. Countermeasures can then be developed to meet the threat.

Identification of Potential or Actual Terrorist Activity

Physical security surveys and inspections, crime prevention surveys, and personal security assessments for high-risk personnel identify existing or potential conditions that may lead to criminal or terrorist activity.

The Installation Vulnerability Determining System

The purpose of the IVDS is to provide a comparative measuring device to assess the vulnerability of an installation. The IVDS, when used in conjunction with physical security, OPSEC evaluations and surveys, and crime prevention programs, provides an accurate assessment of installation vulnerability. Field Manual 19-30 and TC 19-16 outline vulnerability assessment considerations and provide instructions for implementing the IVDS.

THE PHYSICAL SECURITY SURVEY

A physical security survey is a critical on-site examination and analysis of an industrial plant, business, or public or private installation. The purpose of the survey is to ascertain the present security status, to identify deficiencies or excesses, to determine the protection needed, and to make recommendations to improve the overall security.

The Survey Team

The optimum number of members for a team (per single structure) is three: a team chief, a photographer, and a recorder.

The team chief's responsibilities are to—

- Coordinate, schedule, and conduct briefings and interviews.
- Assign responsibilities.
- Supervise.
- Write reports.
- Select photos and photo sequences.
- Conduct inventory of supplies and equipment.
- Write after-action reports.

The photographer selects and procures equipment, takes photographs, marks and develops film, and prints and annotates photos as required.

The recorder is responsible for plans and sketches. He develops symbols and door and lock legends, and records information as directed by the team chief.

8-3

INTELLIGENCE	☐ Identify the local threat via the intelligence gathering process (collection, evaluation, and dissemination of information). ☐ Consider restrictions on collecting and storing information. ☐ Indicate information sources for the intelligence gathering effort (military intelligence, federal agencies, state and local authorities).

THREAT ANALYSIS	☐ Identify local threat (immediate and long-term).
	☐ Identify threats other than local (national and international groups who have targeted or who might target US installations).
	☐ Use IVDS (see page 8-3) when assessing the threat. Consider—
	    • Geography of the area concerned.
	    • Law enforcement resources.
	    • Population factors.
	    • Communication capabilities.
	☐ Establish a priority of identified weaknesses and vulnerabilities.

SECURITY COUNTERMEASURES	☐ Specify terrorist threat conditions and recommended actions. ☐ Include a combination of physical, operations, and personnel security measures.
OPERATIONS SECURITY	☐ Establish procedures to prevent terrorists from readily obtaining plans and operations information (do not publish the commander's itinerary, and safeguard classified material). ☐ Coordinate the installation's OPSEC program. ☐ Include an OPSEC annex in the contingency plan (in accordance with AR 530-1).
	**OPSEC is a vitally important measure that denies information to individuals who will use it for espionage, criminal, or terrorist actions.**
PERSONNEL SECURITY	☐ Identify individuals who are vulnerable to terrorists' attacks. ☐ Point out the identified threats to vulnerable personnel.
PHYSICAL SECURITY	☐ Ensure that special threat plans and physical security plans are mutually supportive. ☐ Establish obstacles to terrorist activity (guards, intrusion detection system, lighting, and fencing). ☐ Include threats identified in DA and MACOM threat statements. ☐ Ensure that the physical security officer assists in threat analysis and corrective action. ☐ Promote command interest in physical security.
AUTHORITY AND JURISDICTION	☐ Indicate in the plan that the FBI has primary investigative and operational responsibility. ☐ Coordinate with the Staff Judge Advocate (SJA). ☐ Allow close cooperation between the principal agents of the military and civilian communities, as well as federal agencies. ☐ Clearly indicate the parameters for the use of force. Brief any elements augmenting military police assets. ☐ Establish an understanding between all local agencies that may be involved in a terrorist incident (military, local FBI resident or senior agent-in-charge, and local law enforcement) regarding authority, jurisdiction, and possible interaction. ☐ Ensure that the SJA has considered the ramifications of closing the post (possible civilian union problems).
CRISIS MANAGEMENT TEAM TRAINING	☐ Establish and exercise the CMT. ☐ Ensure that plans for the CMT are based on needs, recognizing manpower limitations, resource availability, equipment, and command. ☐ Include a location for the CMT. Designate alternate locations and succession of command. ☐ Use visual aids in the emergency operations center or CMT operational areas that provide status reports of situation progress and countermeasures (chalkboards, maps with overlays, and bulletin boards).

Figure 8-2. Special planning considerations.

THREAT MANAGEMENT FORCE TRAINING (TMF headquarters, SRT, and hostage negotiation team)	☐ Train and exercise the TMF under realistic conditions. ☐ Apply corrective action to shortcomings and deficiencies. ☐ Form and mission-specific train the SRT (building entry and search techniques, vehicle assault operations, and antisniper techniques equipment). ☐ Test the SRT quarterly (alert procedures, response time, overall preparedness). ☐ Fix responsibility for the negotiation team (FBI, Criminal Investigation Division [CID], or military police investigator [MPI]). ☐ Train and exercise the negotiation team under realistic conditions. ☐ Properly equip the negotiation team.

Figure 8-2. Special planning considerations (continued).

Data Collecting and Recording

All team members must become familiar with the AO and mission requirements. They collect, record, and report data and coordinate with other team members. They review and disseminate information; develop survey plans and questionnaires; and procure, test, and pack supplies and equipment.

When planning, the survey the team should request information on the AO, the structure to be surveyed, maps, points of contact, and transportation. The team should sterilize the existing plans and sketches and develop a tentative working schedule. The survey plan must be simple, follow a comprehensive sequence, and be flexible.

Team members should coordinate for transportation, identification documents, briefings, interviews, entry accesses to surrounding structures, points of contact, and emergency safe sites. The survey team should ensure that local personnel are aware of its presence in the area. Also, team members should ask about the attitudes of the local population before taking photos. The team chief should ensure that the team uses the appropriate formats, legends, plans, sketches, and equipment.

When collecting physical security information on a structure, team members should begin at the main gate. They should move in a clockwise or counterclockwise direction around the outside perimeter, overlapping the photo coverage. They should collect the gate(s) information, move inside the perimeter, and survey the grounds in the same direction. They should survey the outside of the main structure, starting and ending at the main entrance, always moving in the same direction. (See Figures 8-3 and 8-4.)

When surveying inside the structure, team members should start at the lowest level and work their way up to the roof. They should collect information, as required, about doors, windows, locks, rooms, corridors, stairs, elevators, water points, and fire points. Information on each door, lock, and hinge can be recorded as it is gathered in the plans and sketches. The team may number the doors in the sketch and record the data in a separate door-lock data sheet. The numbering of doors may be continuous from ground-floor level to the roof, or it may be broken down by floors. Every door receives a number. (See Figures 8-5, through 8-8.)

When recording information in the plans and sketches, team members should print in small, legible words. They should use symbols and legends. They should begin the narrative on location, following the security survey outline format (Appendix E) closely. They should also develop film daily by field expedient means if necessary.

8-5

BLOCK FACE AND NEIGHBORHOOD CHARACTERISTICS SURVEY

A block face and neighborhood characteristic survey is a comprehensive collection of data on an area that surrounds a structure to determine the characteristics of the neighborhood as it pertains to security (Figure 8-9). It is accurately reported in a narrative format (see format in Appendix F) that is complete with photographs and sketches.

Use the following techniques when conducting a block face and neighborhood characteristics survey:

- Start at the main gate of the main structure and move in a clockwise or counterclockwise direction out to the boundaries and back to the gate.
- Start at the boundary and move in toward the structure.
- Survey the area, dividing it into cardinal sectors (resembling the cloverleaf reconnaissance technique).
- Follow a figure eight pattern.
- Regardless of the start point, the sequence should end at the main gate.
- The team chief organizes the team, the team reconnoiters the area and adjusts the plan of operation as required.
- Act natural, know the customs, and dress accordingly.
- Avoid or pass quickly through danger areas.
- Be alert to security personnel of other facilities.
- Travel light.
- If in a vehicle, do not drive fast and use a polarizing filter.
- At intersections, using a wide-angle lens, photograph in all cardinal directions to capture on film the street signs for orientation.
- Use ASA 400 film. On hazy days, use an ultraviolet (UV) filter.
- Protect the equipment.
- Know the laws.
- If stopped, explain the nature of the survey.
- If fired at, break contact and seek refuge.
- Use common sense when taking photos.
- Begin the narrative on location. Follow the block face and neighborhood characteristics survey outline format (Appendix F) and write legibly.
- Use symbols and legends for sketches.
- Develop film daily.
- Coordinate and conduct a makeup survey as required and safeguard all information.

PHYSICAL SECURITY EQUIPMENT

The equipment inventory for a physical security survey may include, but is not limited to the items shown in Figure 8-10.

8-6

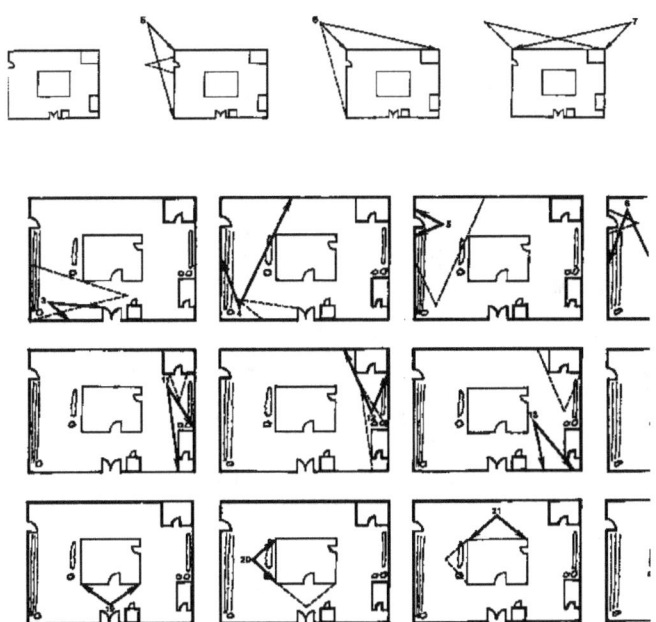

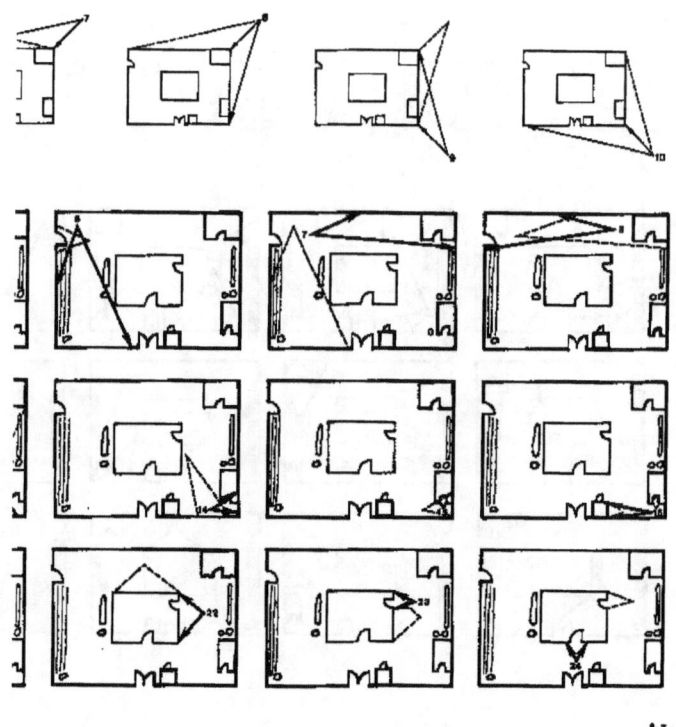

GENERAL	☐ Survey formats (physical security, block face and neighborhood characteristics, checklist) ☐ Maps, charts, military maps (1:50,000; 1:25,000; 1:12,500), city maps, aerial photographs ☐ Blueprints, floor plans, site plans, or other available sketches ☐ Tape recorder with batteries ☐ Penlight, flashlight ☐ Briefcase, clipboards ☐ Manila folders and expanding folders (to organize compiled information) ☐ Notebooks, paper pads, graph paper ☐ Pens, pencils, felt tip pens and markers, erasers, pencil sharpener ☐ Rulers (inches and metric), measuring tape (as required) ☐ Protractor and compass ☐ Templates (circles, squares, and French curves) ☐ Gum labels ☐ Cellophane and masking tape (1-inch and 2-inch widths) ☐ Trash bags
PHOTOGRAPHIC	☐ Camera body (minimum of two) ☐ Camera lenses (28 mm and 35 mm wide-angle, 50 mm standard, 35 mm telephoto) ☐ Camera filters (UV, polarizing) ☐ Camera flash ☐ Tripod, cable release ☐ Extra batteries (for camera and flash) ☐ Camera cleaning equipment ☐ Developing guide ☐ Dark bag ☐ Camera bag ☐ Photographic chemicals (preferably dry chemicals, Microdol, D-76 [to push film], fix, and wetting solution) ☐ Bottles for chemicals ☐ Canister, reels ☐ Plastic aprons ☐ Thermometer, funnel, measuring cup, bottle opener ☐ Scissors ☐ Negative holders ☐ Film (preferably in bulk rolls; Tri-X Pan, Plus-X Pan, Ektachrome [ASA 64 and 400]) ☐ Bulk film rollers (as required) ☐ 35-mm canisters (for bulk film loading)

Figure 8-10. Sample physical security equipment list.

APPENDIX A

AMBUSH FORMATIONS

POINT AMBUSH

A point ambush, whether independent or part of an area ambush, is positioned along the target's expected route of approach. Formation is important because, to a great extent, it determines whether a point ambush can deliver the heavy volume of highly concentrated fire necessary to isolate, trap, and destroy the target.

The formation to be used is determined by carefully considering possible formations and the advantages and disadvantages of each in relation to terrain, conditions of visibility, forces, weapons and equipment, ease or difficulty of control, target to be attacked, and overall combat situation.

This appendix discusses a few formations that have been developed for the deployment of point ambushes. Those discussed are named according to the general pattern formed on the ground by the deployment of the attack element.

Point Ambush Formations

Line. The attack element is deployed generally parallel to the target's route of movement (road, trail, stream). This deployment positions the attack element parallel to the long axis of the killing zone and subjects the target to heavy flanking fire. The size of the target, which can be trapped in the killing zone, is limited by the area the attack element can effectively cover with a heavy volume of highly concentrated fire. The target is trapped in the killing zone by natural obstacles, mines (Claymore, antivehicular, antipersonnel), demolitions, and direct and indirect fires (Figure A-1). A disadvantage of the line formation is the chance that lateral dispersion of the target may be too great for effective coverage. Line formation is appropriate in close terrain that restricts target maneuver and in open terrain where one flank is restricted by mines, demolitions, mantraps, or sharpened stakes. Similar obstacles can be placed between the attack element and the killing zone to provide protection from the target's counterambush measures. When a destruction ambush is deployed in this manner, access lanes are left so that the target can be assaulted (Figure A-2). The line formation can be effectively used by a *rise from the ground* ambush in terrain seemingly unsuitable for ambush. An advantage of the line formation is its relative ease of control under all conditions of visibility.

The L. The L-shaped formation is a variation of the line formation. The long side of the attack element is parallel to the killing zone and delivers flanking fire. The short side of the attack element is at the end of and at right angles to the killing zone and delivers enfilading fire that interlocks with fire from the other leg. This formation is very flexible. It can be established on a straight stretch of a trail or stream (Figure A-3), or at a sharp bend in a trail or stream (Figure A-4). When appropriate, fire from the short leg can be shifted to parallel the long leg if the target attempts to assault or escape in the opposite direction. In addition, the short leg prevents escape in the direction of attack element and reinforcement from its direction (Figure A-5).

The Z. The Z-shaped formation is another variation of the line formation. The attack force is deployed as in the L formation, but with an additional side so that the formation resembles the letter Z. The additional side (Figure A-6) may serve—

A-1

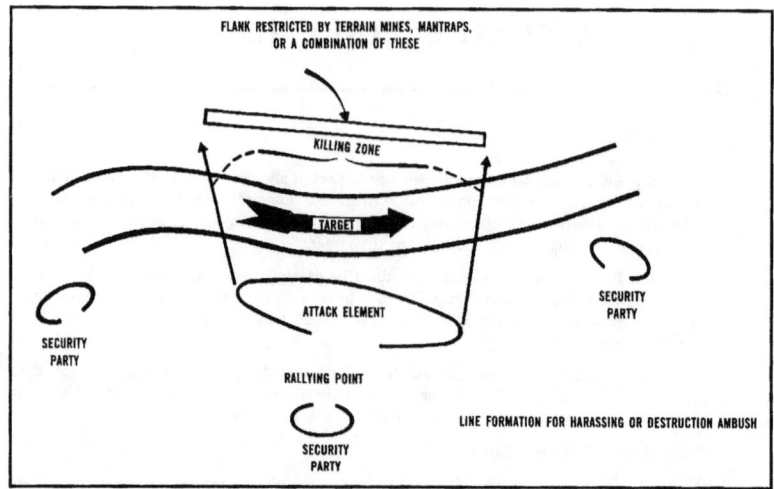

Figure A-1. Line formation, example 1.

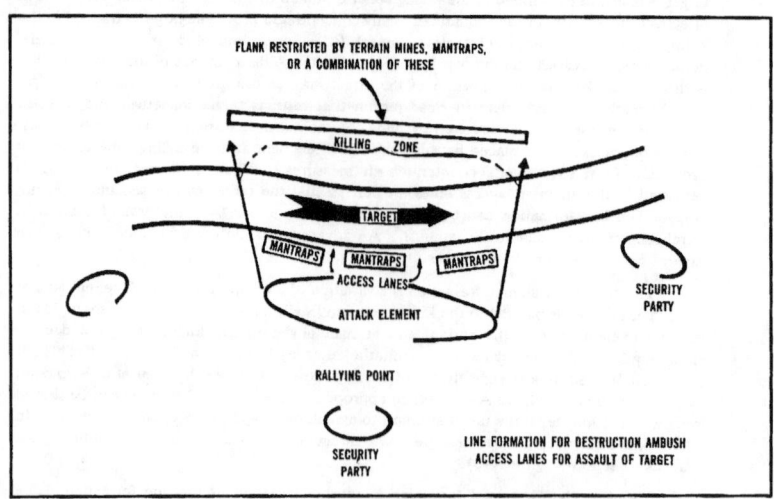

Figure A-2. Line formation, example 2.

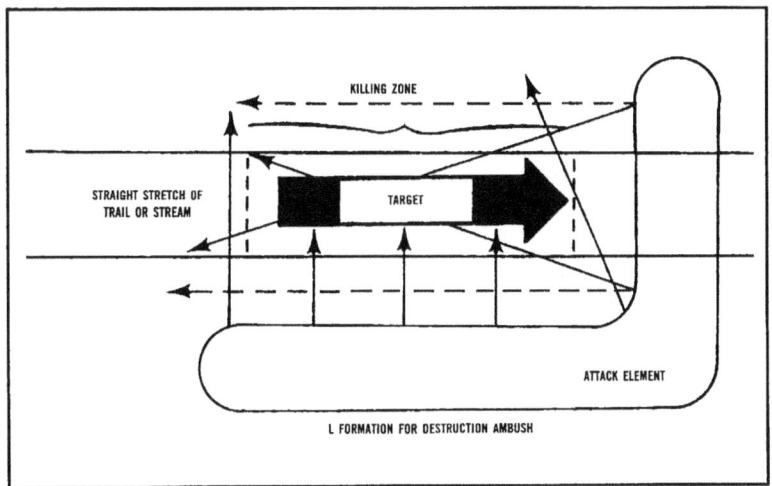

Figure A-3. L formation, example 1.

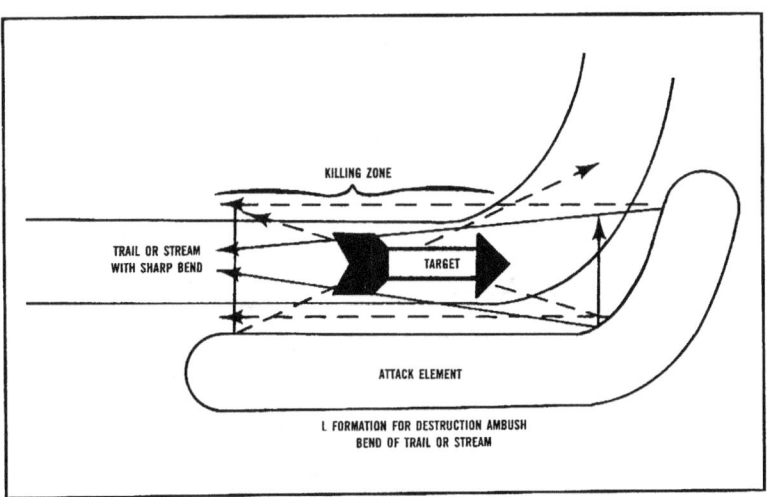

Figure A-4. L formation, example 2.

A-3

- To engage a force attempting to relieve or reinforce the target.
- To seal the end of the killing zone.
- To restrict a flank.
- To prevent envelopment.

The T. In the T–shaped formation, the attack element is deployed across and at right angles to the target's route of movement so that it and the target form the letter T. This formation can be used day or night to establish a purely harassing ambush and at night to establish an ambush to interdict movement through open, hard–to–seal areas (such as rice paddies).

A small group of persons can use the T formation to harass, slow, and disorganize a larger force. When the lead elements of the target are engaged, they will normally attempt to maneuver right or left to close with the ambush. Mines, mantraps, and other obstacles placed to the flanks of the killing zone slow the enemy's movements and permit the ambush patrol to deliver heavy fire and withdraw without becoming decisively engaged (Figure A–7).

The T formation can be used to interdict small groups attempting night movement across open areas. For example, the attack element is deployed along a rice paddy dike with every second person facing in the opposite direction. The attack of a target approaching from either direction requires only that every second person shift to the opposite side of the dike. Each person fires only to his front and only when the target is at very close range. Attack is by fire only and each person keeps the target under fire as long as it remains on his front. If the target attempts to escape in either direction along the dike, each man takes it under fire as it comes to his vicinity. The T formation is very effective at halting infiltration. But it has one chief disadvantage: while spread out, the ambush may engage a superior force. Use of this formation must, therefore, fit the local enemy situation (Figure A–8).

The V. The V–shaped attack element is deployed along both sides of the target's route of movement so that it forms the letter V; care is taken to ensure that neither group (nor leg) fires into the other. This formation subjects the target to both enfilading and interlocking fire. The V formation is best suited for fairly open terrain but can also be used in the jungle. When established in the jungle, the legs of the V close in as the head elements of the target approach the apex of the V; the attack element then opens fire from close

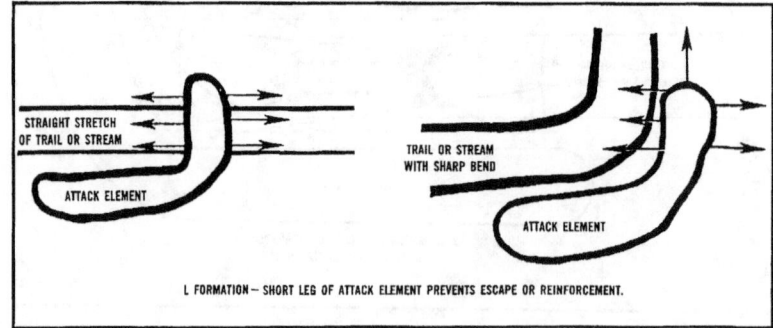

L FORMATION— SHORT LEG OF ATTACK ELEMENT PREVENTS ESCAPE OR REINFORCEMENT.

Figure A–5. L formation.

A–4

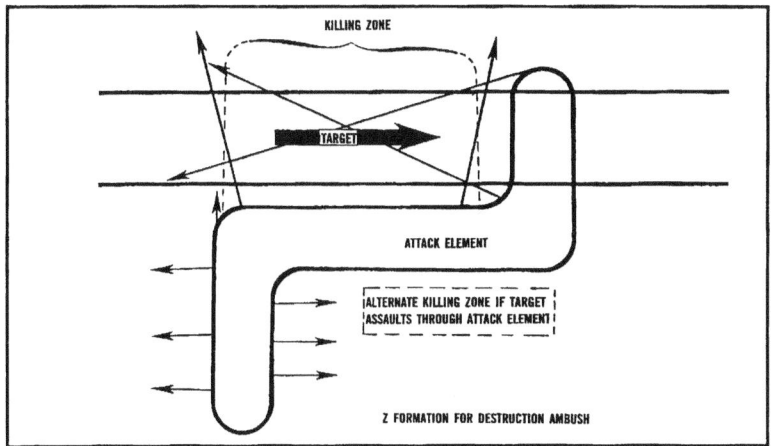

Figure A-6. Z formation.

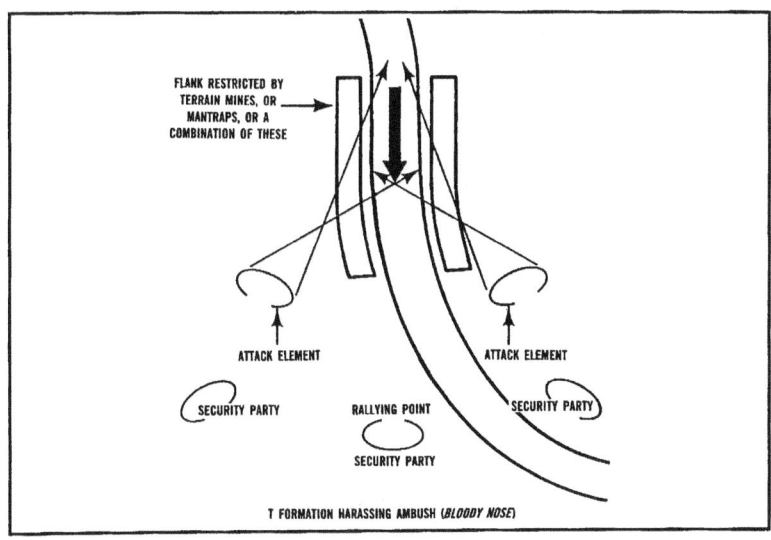

Figure A-7. T formation, example 1.

171

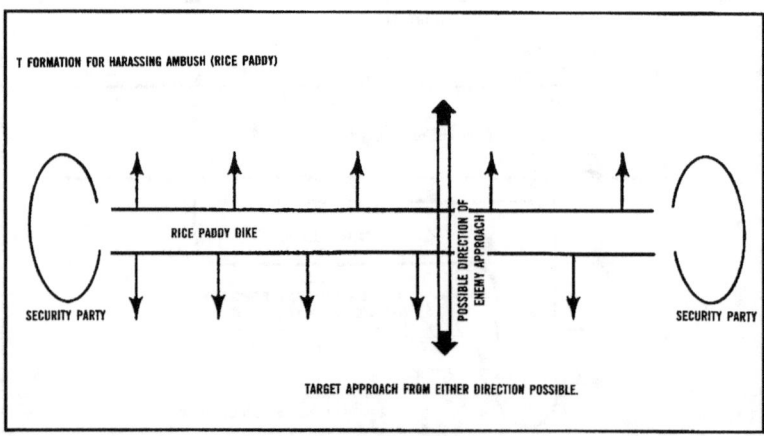

T FORMATION FOR HARASSING AMBUSH (RICE PADDY)

RICE PADDY DIKE

POSSIBLE DIRECTION OF ENEMY APPROACH

SECURITY PARTY

SECURITY PARTY

TARGET APPROACH FROM EITHER DIRECTION POSSIBLE.

Figure A-8. T formation, example 2.

range. Here, even more than in open terrain, all movement and fire must be carefully coordinated and controlled to ensure that the fire of one leg does not endanger the other. The wider separation of elements makes this formation difficult to control, and there are fewer sites that favor its use. Its main advantage is that it is difficult for the target to detect the ambush until it has moved well into the killing zone (Figures A-9 and A-10).

Triangle. This is a variation of the V formation and can be varied in three ways.

In a *closed triangle* (Figure A-11), the attack element is deployed in three groups or parties, positioned so that they form a triangle (or closed V). An automatic weapon is placed at each point of the triangle and positioned so that it can be shifted quickly to interlock with either of the others. Men are positioned so that their fields of fire overlap. Mortars may be positioned inside the triangle. When deployed in this manner, the triangle ambush becomes a small unit strongpoint. It is used to interdict night movement through rice paddies and other open areas when target approach is likely to be from any direction. The formation provides all-around security, and security parties are deployed only when they can be positioned so that if detected by an approaching target, they will not compromise the ambush. Attack is by fire only, and the target is allowed to approach within close range before fire is opened.

Advantages include ease of control and all-around security. In addition, a target approaching from any direction can be brought under fire of at least two automatic weapons.

There are several disadvantages. For example, an ambush patrol of platoon size or larger is required to reduce the danger of being overrun by an unexpectedly large target. One or more legs of the triangle may come under enfilade fire. Lack of dispersion, particularly at the points, increases danger from enemy mortar fire.

The *open triangle* (*harassing ambush*), a variation of the triangle ambush, is designed to enable a small force to harass, slow, and inflict heavy casualties upon a larger force without

A-6

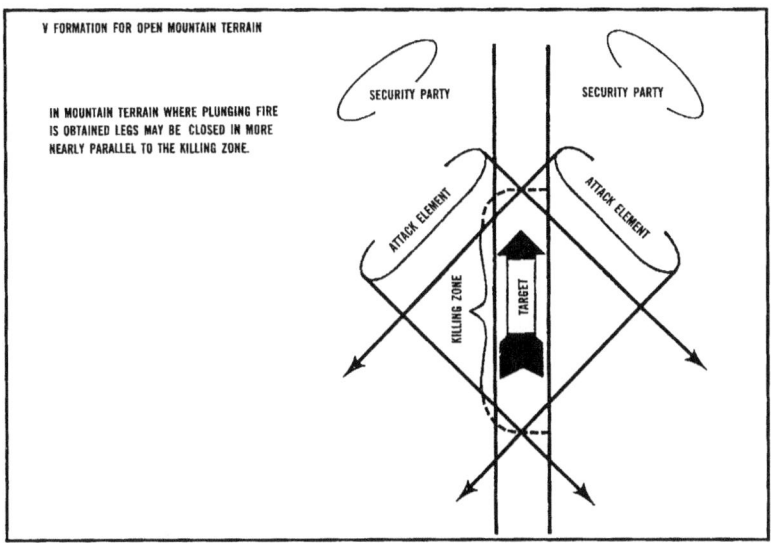

Figure A-9. V formation, example 1.

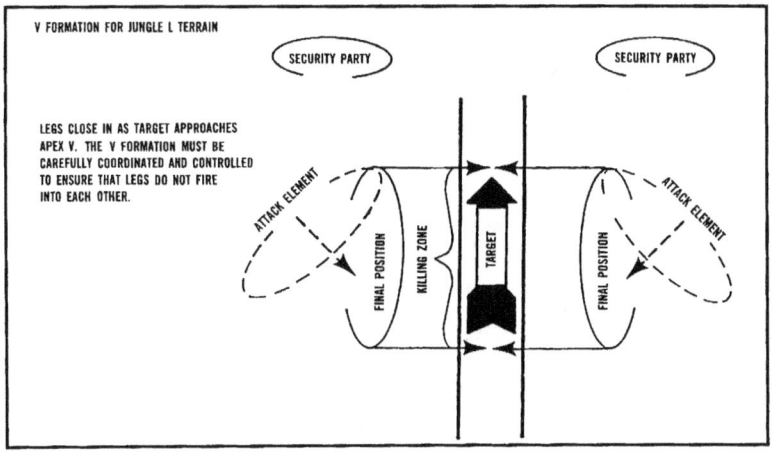

Figure A-10. V formation, example 2.

A-7

173

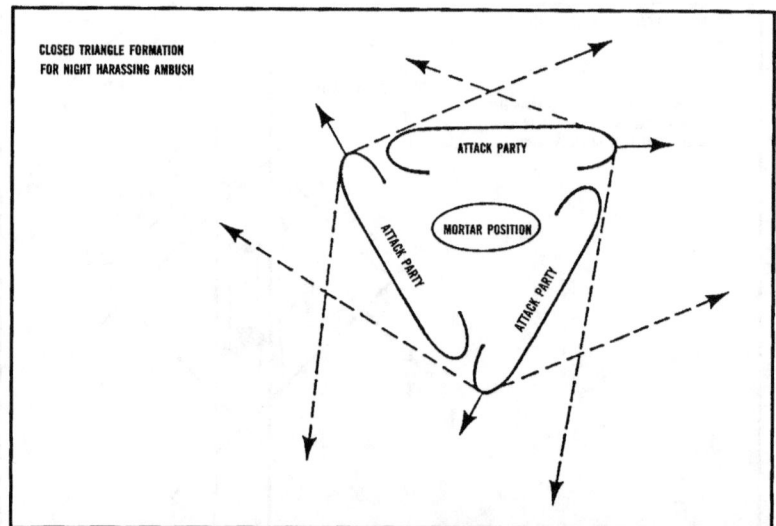

CLOSED TRIANGLE FORMATION
FOR NIGHT HARASSING AMBUSH

ATTACK PARTY

MORTAR POSITION

ATTACK PARTY

ATTACK PARTY

Figure A-11. Closed triangle formation.

itself being decisively engaged. The attack element is deployed in three parties, positioned so that each party becomes a corner of a triangle containing the killing zone. When the target enters the killing zone, the party to the target's front opens fire on the leading element. When the target counterattacks, the group withdraws and an assault party to the flank opens fire. When this party is attacked, the party to the opposite flank opens fire. This process is repeated until the target is pulled apart. Each party reoccupies its position, if possible, and continues to inflict the maximum damage possible without becoming decisively engaged (Figure A-12).

In an *open triangle (destruction ambush)*, the attack element is again deployed in three parties, positioned so that each party is a point of the triangle, 200 to 300 meters apart. The killing zone is the area within the triangle. The target is allowed to enter the killing zone; the nearest party attacks by fire. As the target attempts to maneuver or withdraw, the other groups open fire. One or more assault parties, as directed, assault or maneuver if possible, envelop or destroy the target (Figure A-13). As a destruction ambush, this formation is suitable for platoon-size or larger forces. A unit smaller than a platoon would be in too great a danger of being overrun.

The following are more disadvantages of the triangle:

- In assaulting or maneuvering, control is very difficult. Very close coordination and control are necessary to ensure that assaulting or maneuvering assault parties are not fired on by another party.

A-8

174

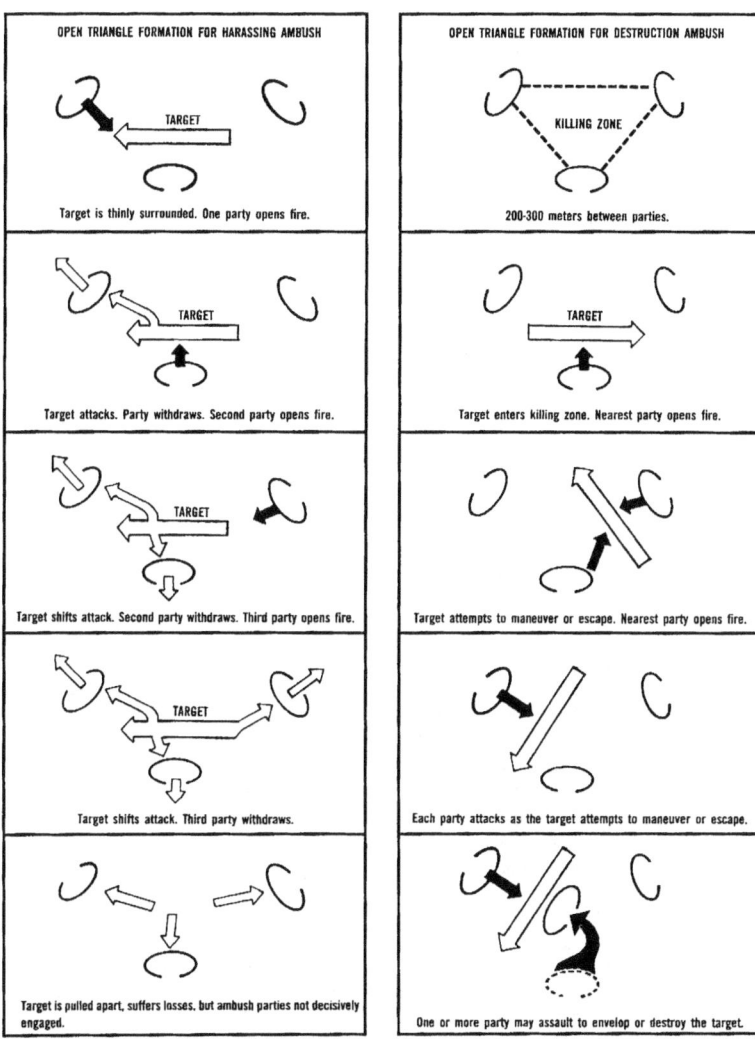

OPEN TRIANGLE FORMATION FOR HARASSING AMBUSH	OPEN TRIANGLE FORMATION FOR DESTRUCTION AMBUSH
Target is thinly surrounded. One party opens fire.	200-300 meters between parties.
Target attacks. Party withdraws. Second party opens fire.	Target enters killing zone. Nearest party opens fire.
Target shifts attack. Second party withdraws. Third party opens fire.	Target attempts to maneuver or escape. Nearest party opens fire.
Target shifts attack. Third party withdraws.	Each party attacks as the target attempts to maneuver or escape.
Target is pulled apart, suffers losses, but ambush parties not decisively engaged.	One or more party may assault to envelop or destroy the target.

Figure A-12. Open triangle formation, example 1.

Figure A-13. Open triangle formation example 2.

A-9

175

- The ambush site must be a fairly level, open area that provides (around its border) concealment for the ambush patrol (unless it is a *rise from the ground* ambush).

Box. This formation is similar in purpose to the open triangle ambush. The attack element is deployed in four parties, positioned so that each party becomes a corner of a square or rectangle containing the killing zone. It can be used as a harassing or destruction ambush in the same manner as the two variations of the open triangle ambush (Figures A-14 and A-15).

AREA AMBUSH

The origin of the type of ambush now called *area ambush* is not known. It was used by Hannibal against the Romans in the second century B.C. More recently, it was modified and perfected by the British Army in Malaya and, with several variations, used in Vietnam. The British found that point ambushes often failed to produce heavy casualties. When ambushed, the Communist guerrillas would immediately break contact and disperse along escape routes leading away from the killing zone. The British counteracted this tactic by blocking escape routes with point ambushes. They called these multiple *related* point ambushes the area ambush.

British Version

The British Army version of the area ambush is as follows:

- A point ambush is established at a site having several trails or other escape routes leading away from it. The site may be a water hole, an enemy campsite, a known rendezvous point, or along a frequently traveled trail. This site is the central killing zone.
- Point ambushes are established along the trails or other escape routes leading away from the central killing zone.
- The target, whether a single group or several groups approaching from different directions, is permitted to move to the central killing zone. Outlying ambushes do not attack (unless discovered).
- The ambush is initiated when the target moves into the central killing zone.
- When the target breaks contact and attempts to disperse, escaping portions are intercepted and destroyed by the outlying ambushes.
- The multiple contacts achieve increased casualties, harassment, and confusion (Figure A-16).

The British Army version of the area ambush is best suited to counterguerrilla operations in terrain where movement is largely restricted to trails. It produces the best results when it is established as a deliberate ambush.

When there is not sufficient intelligence for a deliberate ambush, an area ambush of opportunity may be established. The outlying ambushes are permitted to attack targets approaching the central killing zone, if within their capability. If too large for the particular outlying ambush, the target is allowed to continue and is attacked in the central killing zone.

Baited Trap Version

A variation of the area ambush is the baited trap version (Figure A-17), where a central killing zone is established along the target's route of approach.

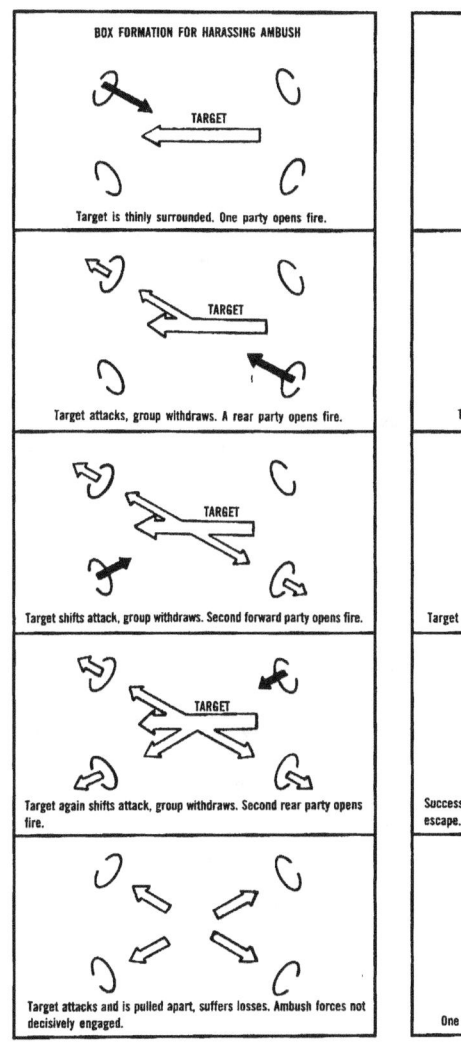

Figure A-14. Box formation,
example 1.

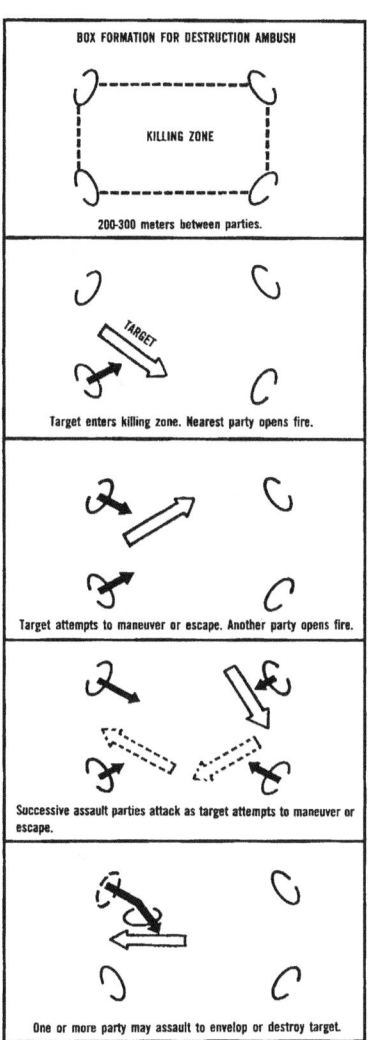

Figure A-15. Box formation.
example 2.

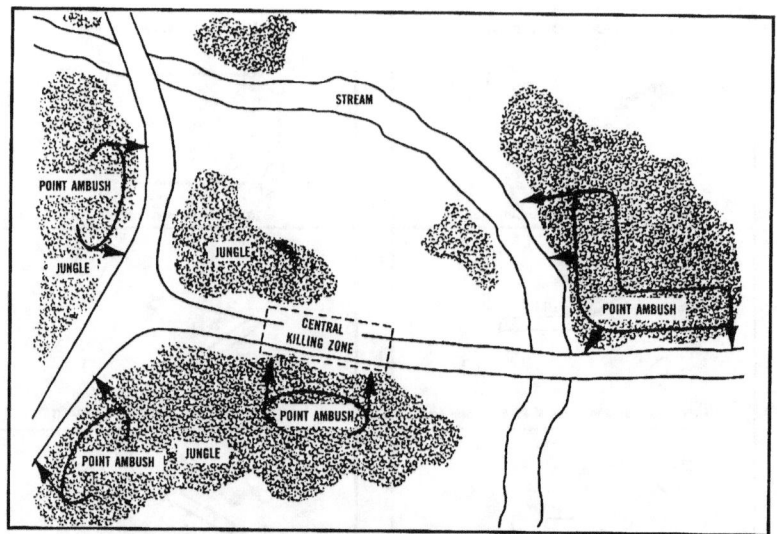

Figure A-16. Area ambush, British version.

Point ambushes are established along the routes over which relieving or reinforcing units will have to approach.

The target in the central killing zone serves as *bait* to lure relieving or reinforcing units into the killing zones of the outlying ambushes.

The outlying point ambushes need not be strong enough to destroy their targets. They may be small, harassing ambushes that delay, disorganize, and *eat away* the target by successive contacts.

This version can be varied by using a fixed installation as *bait* to lure relieving or reinforcing units into the killing zone of one or more of the outlying ambushes. The installation replaces the central killing zone and is attacked. The attack may intend to overcome the installation or may be only a ruse.

These two variations are best suited for situations where routes of approach for relieving or reinforcing units are limited to those favorable for ambush.

They are also best suited for use by guerrilla forces, rather than counterguerrilla forces. Both variations were used extensively by Communist guerrilla forces in Vietnam.

UNUSUAL AMBUSH TECHNIQUES

The ambush techniques described above are so well known and widely used that they are considered *standard*. Other, less well known, less frequently used techniques are considered *unusual*. Two such techniques are described on page A-13.

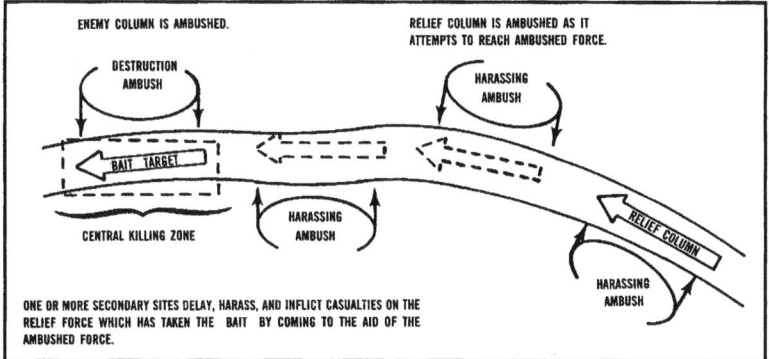

Figure A-17. Area ambush, *baited trap* version.

Rise from the Ground Ambush

This point ambush is designed (Figure A-18) for use in open areas that lack the good cover and concealment and other features normally desirable in a *good* ambush site. The attack element is deployed in the formation best suited to the overall situation.

The attack element is completely concealed in the *spider hole* type of covered foxhole. Soil is carefully removed and positions expertly camouflaged.

When the ambush begins, the attack element throws back the covers and literally *rises from the ground* to attack.

This ambush takes advantage of the tendency of patrols and other units to relax in areas that do not appear to favor ambush.

The chief disadvantage is that the ambush patrol is very vulnerable if prematurely detected.

Demolition Ambush

Electrically detonated mines or demolition charges, or both, are planted in an area (Figure A-19) over which a target is expected to pass. This may be a portion of a road or trail, an open field, or any area that can be observed from a distance. Activating wires are run to a concealed observation point, which is sufficiently distant to ensure safety of the ambushers.

As large a force as desired or necessary can be used to mine the area. Two men remain to begin the ambush; others return to the unit.

When a target enters the mined area (killing zone), the two men remaining detonate the explosives and withdraw immediately to avoid detection and pursuit.

A-13

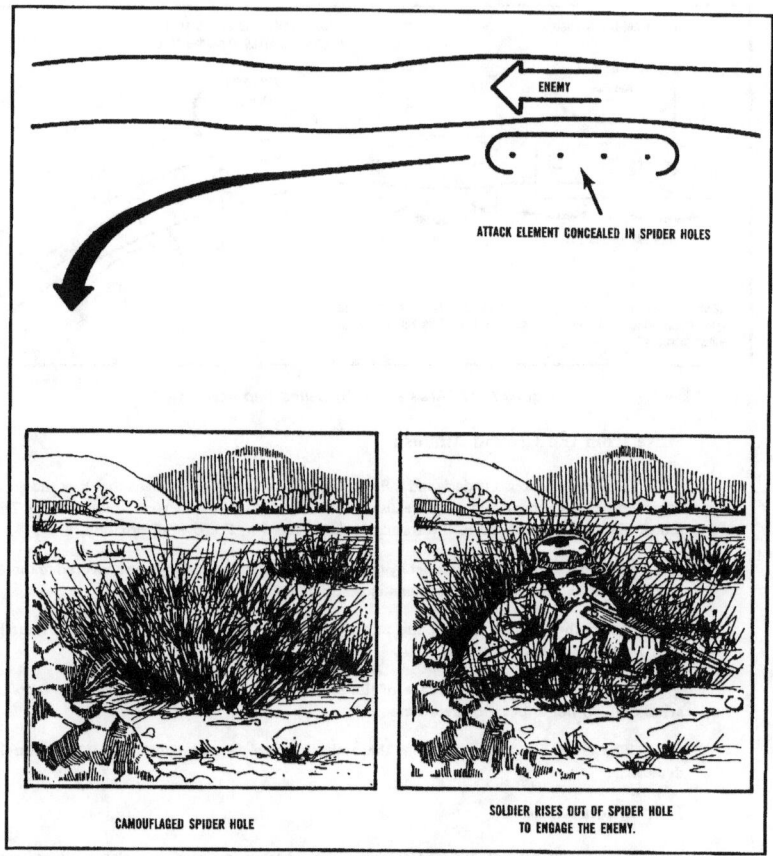

Figure A-18. Unusual techniques, *rise from the ground* ambush.

Special Ambush Situations

Columns protected by armor. Attacks against columns protected by armored vehicles depend on the type and location of armored vehicles in a column and the weapons of the ambush patrol. If possible, armored vehicles are destroyed or disabled by fire of antitank weapons, landmines, Molotov cocktails, or by throwing hand grenades into open hatches. An effort is made to immobilize armored vehicles at a point where they are unable to give protection to the rest of the convoy and where they will block the route of other supporting vehicles.

A-14

Figure A–19. Unusual techniques, demolition ambush.

Ambush of trains. Moving trains may be subjected to harassing fire, but the most effective ambush is derailment. Train derailment is desirable because the wreckage remains on the tracks and delays traffic for long periods of time. Derailment on a grade, at a sharp curve, or on a high bridge will cause most of the cars to overturn and result in extensive casualties among the passengers. Fire is directed on the exits of overturned coaches; and designated parties, armed with automatic weapons, rush forward to assault coaches or cars still standing. Other parties take supplies from freight yards and then set fire to the train. Rails are removed from the track at some distance from the ambush site in each direction to delay the arrival of reinforcements by train. In planning the ambush of a train, remember that the enemy may include armored railroad cars in the train for its protection and that important trains may be preceded by advance guard locomotives or inspection cars to check the track.

Ambush of waterway traffic. Waterway traffic, such as barges or ships, may be ambushed similar to a vehicular column. The ambush patrol may be able to mine the waterway and thus stop traffic. If mining is not feasible, fire delivered by recoilless weapons can damage or sink the craft. Fire should be directed at engine room spaces, the waterline, and the bridge. Recovery of supplies may be possible if the craft is beached on the banks of the waterway or grounded in shallow water.

AMBUSH PATROLS

An *ambush patrol* is a combat patrol whose mission is to—

- Harass a target.
- Destroy a target.
- Capture personnel or equipment.
- Execute any combination of these.

An ambush patrol is planned and prepared in the same general manner as other patrols, that is, by using patrol steps (troop leading procedures).

Planning and Preparation

Planners must first consider whether the ambush is to be a deliberate ambush or an ambush of opportunity. In a deliberate ambush, the greater amount of target intelligence available permits planning for every course of action at the target. Plans for an ambush of opportunity must include consideration of the types of targets that may be ambushed, as well as varying situations. In both, plans must be flexible enough to allow modification, as appropriate, at the ambush site. When planning, the principles discussed below apply. All plans must be rehearsed in detail.

Simplicity. Every person must thoroughly understand what he is to do at every stage of the operation. In ambush, more so than in other operations, failure of even one person to perform exactly as planned can cause failure.

Type of ambush. The type of ambush (*point* or *area*) affects the organization, the number of men required, the equipment and communications required, and all other aspects of the patrol.

Deployment. Each possible formation must be considered for its advantages and disadvantages.

Manner of attack. An attack may be by fire only (*harassing ambush*) or may include an assault of the target (*destruction ambush*).

Size of ambush patrol. The patrol is tailored for its mission. Two men may be adequate for an harassing ambush. A destruction ambush may require the entire unit (squad, platoon, company).

Organization. An ambush patrol is organized in the same manner as other combat patrols to include a patrol headquarters, an assault element, a support element, and a security element (Figure A-20). The assault and support elements are the attack force; the security element is the security force. When appropriate, the attack force is further organized to provide a reserve force. When an ambush site is to be occupied for an extended period, double ambush patrols may be organized. One ambush patrol occupies the site while the other rests, eats, and tends to personal needs at the ORP or other concealed location. They alternate each 8 hours. If the waiting period is over 24 hours, three ambush patrols are organized (Figure A-21).

Equipment. The selection of accompanying equipment and supplies is based on the—
- Mission.
- Enemy threat.
- Size of the resistance force.
- Means of transportation.
- Distance and terrain.
- Weight and bulk of equipment.

Routes. A primary route is planned that will allow the patrol to enter the ambush site from the rear. The killing zone is not entered if entry can be avoided. If the killing zone

A-16

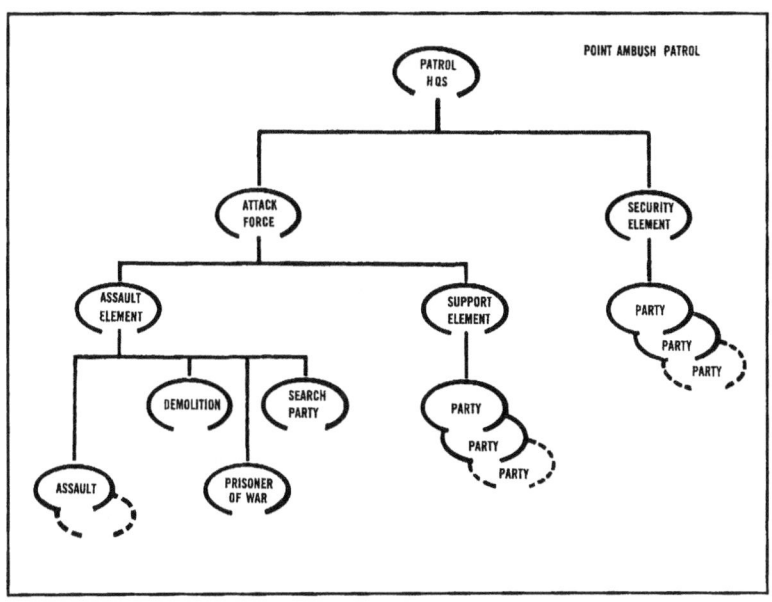

Figure A-20. Organization of ambush patrols, example 1.

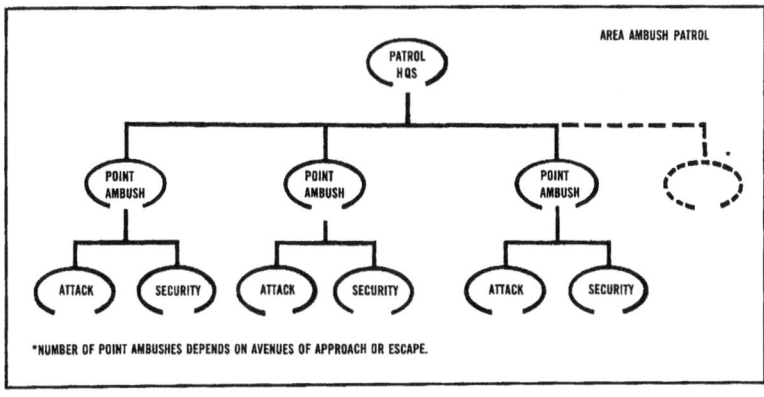

Figure A-21. Organization of ambush patrols, example 2.

A-17

183

must be entered to place mines or explosives, great care must be taken to remove any tracks and signs that might alert the target and compromise the ambush. If mines, mantraps, or explosives are to be placed on the far side, or if the appearance of the site from the target's viewpoint is to be checked, a wide detour around the killing zone is made. Here, too, great care must be taken to remove any traces that might reveal the ambush.

An alternate route from the ambush site is planned, as in other patrols.

Site. Maps and aerial photographs are used to analyze the terrain. When possible, the patrol makes an on-the-ground reconnaissance. As far as possible, so-called *ideal* ambush sites are avoided. An alert enemy is suspicious of these areas, avoids them if possible, and increases vigilance and security when they must be entered. Surprise is even more difficult to achieve in these areas. Instead, apparently unlikely sites are chosen when possible. Considering this, an ambush site must provide—

- Favorable fields of fire.
- For occupation and preparation of concealed positions.
- Canalization of the target into the killing zone.
- Covered routes of withdrawal to enable the ambush patrol to break contact and avoid pursuit by effective fire.

Occupation of the site. As a general rule, the ambush patrol occupies the ambush site at the latest possible time permitted by the tactical situation and by the amount of time required to prepare the site preparation required. This not only reduces the risk of discovery but also reduces the time men must remain still and quiet in position.

Positions. The patrol moves into the ambush site from the rear as discussed earlier. Security elements are positioned first to prevent surprise while the ambush is being established. Automatic weapons are then positioned so that each can fire along the entire killing zone. If this is not possible, they are given overlapping sectors of fire so that the entire killing zone is covered. The patrol leader then selects his position, located where he can tell when to begin the ambush. Riflemen and grenadiers are then placed to cover any dead space left by the automatic weapons. All weapons are assigned sectors of fire to provide mutual support. The patrol leader sets the position preparation time. The degree of preparation depends on the time allowed. All men work at top speed during the allotted time.

Camouflage. Camouflage is of utmost importance. Each man must be hidden from the target. During preparation for the patrol, each man camouflages himself and his equipment and secures his equipment to prevent noise. At the ambush site, positions are prepared with minimum change in the natural appearance of the site. All debris resulting from preparation of positions is concealed.

Execution

Signals. Three signals, often four, are needed to execute the ambush. Audible and visual signals, such as whistles and pyrotechnics, must be changed often to avoid establishing patterns. Too frequently, use of the same signals may result in their becoming known to the enemy. A target might recognize a signal and be able to react in time to avoid the full effects of the ambush. For example, if a white star cluster is habitually used to signal withdrawal in a night ambush, an alert enemy might fire one and cause premature withdrawal.

A signal by the security force to alert the patrol leader to the target's approach may be given by—

- Arm-and-hand signals.

- Radio (as a *quiet* voice message), by transmitting a prearranged number of taps, or by signaling with the push-to-talk switch.

- Field telephone, when there is no danger that wire between positions will compromise the ambush.

A signal to begin the ambush, given by the patrol leader or a designated individual, may be a shot or the detonation of mines or explosives.

A signal for lifting or shifting fires, if the target is to be assaulted, may be given by voice command, whistles, or pyrotechnics. All fire must stop immediately so that the assault can be made before the target can react.

A signal for withdrawal may also be by voice command, whistles, or pyrotechnics.

Fire discipline. This is a key part of the ambush. Fire must be withheld until the signal is given, then immediately delivered in the heaviest, most accurate volume possible. Properly timed and delivered fires achieve surprise as well as destruction of the target. When the target is to be assaulted, the lifting or shifting of fires must be equally precise. Otherwise, the assault is delayed and the target has opportunity to recover and react.

Withdrawal to the ORP. The ORP is located far enough from the ambush site that it will not be overrun if the target attacks the ambush. Routes of withdrawal to the ORP are reconnoitered. Situation permitting, each person walks the route he is to use and picks out checkpoints. When the ambush is executed at night, each person must be able to follow his route in the dark.

On signal, the patrol quickly but quietly withdraws to the ORP, reorganizes, and begins its return march.

If the ambush was not successful and the patrol is pursued, withdrawal may be by bounds. The last group may arm mines, previously placed along the withdrawal route, to further delay pursuit.

Contingency plans should include removal of the wounded, both friendly and hostile, under pursuit or at a more measured pace. Treatment location and moves from the target site to a rearward position must be flexible. Plans should also include insertion of medical assets within the assault element, as well as within the headquarters. Security and support elements should be considered, depending on the mission.

APPENDIX B

IMMEDIATE ACTION DRILLS

A patrol may make contact with the enemy at any time. This is especially true in guerrilla operations. Contact may be by chance, by air observation or attack, or by ambush. It may be visual only: The patrol sights the enemy but is undetected by it. In this case, the patrol leader can decide whether to make or avoid physical contact, basing his decision on the patrol's mission and capability to successfully engage the potential target.

When a patrol's mission prohibits physical contact, except that necessary to accomplish the mission, its actions are defensive in nature. It avoids being seen by the enemy. Physical contact, if unavoidable, is broken as quickly as possible, and the patrol, if still capable, continues its mission.

When a patrol's mission permits or requires it to seek or exploit opportunities for contact (as in the case of search and attack patrols), its actions are offensive, immediate, and position.

In foot patrolling, especially in guerrilla operations, contacts (visual or physical) are often unexpected, at very close range, and of short duration. Effective fire, or the threat of fire, often gives leaders little or no time to fully estimate situations and issue orders. In these situations, immediate action drills provide a means for swiftly starting positive offensive or defensive action, as appropriate.

The immediate action drills discussed below are drills designed to provide swift and positive small unit reaction to enemy visual or physical contact. They are simple courses of action in which all men are so well trained that minimum signals or commands are required to start action.

IMMEDIATE HALT DRILL

When the situation requires the immediate, in-place halt of the patrol, the immediate action drill *freeze* is used. This is the situation when the patrol detects the enemy but is undetected by it. The first man detecting the enemy (visually or otherwise) gives the special silent signal, *freeze*. Every man halts *in place*, weapon at the ready, and remains absolutely motionless and quiet until further signals or orders are given.

AIR OBSERVATION AND ATTACK DRILLS

These drills are designed to reduce the danger of detection by aircraft and casualties from low-level air attack.

Air Observation

When an aircraft (enemy or unidentified) that may detect the patrol is heard or observed, the appropriate immediate action drill is *freeze*. The first man hearing or sighting such an aircraft signals *freeze*. Every man freezes in place until the patrol leader identifies the aircraft and gives further signals or orders.

B-1

187

Air Attack

When an aircraft detects a patrol and makes a low–level attack, the immediate action drill *air attack* is used. The first man sighting an attacking aircraft shouts, "aircraft front (left, rear, or right)." The patrol moves quickly into line formation, well spread out, *at right angles* to the aircraft's direction of travel. As each man comes on line, he hits the ground, using available cover. He positions his body *at right angles* to the aircraft's direction of travel to present the shallowest target possible.

Between attacks (if the aircraft returns or if more than one aircraft attacks) men seek better cover.

Attacking aircraft are fired on only on command of the patrol leader.

CHANCE CONTACT DRILLS

Hasty Ambush

This immediate action drill is a defensive measure used to avoid contact and an offensive measure used to make contact. It may often be a subsequent action, *freeze*. When the special silent signal *hasty ambush* is given (by a point man, a patrol leader, or another authorized person), the entire patrol moves quickly to the right or left of line of movement, as indicated by signal, and takes up the best available concealed firing positions. The patrol leader begins the ambush by opening fire and shouting, "fire." This ensures the start of the ambush if his weapon misfires. If the patrol is detected before this, the first person aware of detection begins the ambush by firing and shouting.

When used as a defensive measure to avoid contact, the ambush is not started unless the team is detected.

When used as an offensive measure, the enemy is allowed to advance until it is in the most vulnerable position before beginning the ambush.

An alternate means for beginning the ambush is to designate an individual (for example, point or last man) to open fire when a certain portion of the enemy reaches or passes him.

Immediate Assault

This immediate action drill is used defensively to make and quickly break unavoidable contact (including ambush). It is also used offensively to decisively engage the enemy (including ambush). When used in chance contact, men nearest the enemy open fire and shout, "contact, front (right, left, or rear)." The patrol moves swiftly into line formation and assaults.

When used defensively, the assault is stopped if the enemy withdraws and contact is broken quickly. If the enemy stands fast, the assault is carried through enemy positions and movement is continued until contact is broken.

When used offensively, the enemy is decisively engaged. Anyone attempting to escape is pursued and destroyed.

COUNTERAMBUSH DRILLS

When a patrol is ambushed, the immediate action drill is determined by nearness or farness of the ambush. (See Chapter 1 for discussion of *near* and *far* ambushes).

In a *near* ambush, the killing zone is under very heavy, highly concentrated, close-range fires. There is little time or space for patrol members to maneuver or seek cover. The longer they remain in the killing zone, the more certain their destruction. Therefore, if attacked by a *near* ambush, react as follows:

- Men in the killing zone, *without order or signal*, immediately assault directly into the ambush position, occupy it, and continue the attack or break contact, as directed. This action moves the patrol members out of the killing zone, prevents other elements of the ambush from firing on them without firing on their own men, and provides positions from which other actions may be taken.

- Men not in the killing zone maneuver, as directed, against the attack force and other elements of the ambush.

- The attack is continued to eliminate the ambush or to break contact as directed.

In a *far* ambush, the killing zone is also under very heavy, highly concentrated fires, but from a greater range. This greater range provides patrol members in the killing zone some space for maneuver and some opportunity to seek cover at a lesser risk of destruction. Therefore, if attacked by a *far* ambush, react as follows:

- Men in the killing zone, without order or signal, immediately return fire, take the best available positions, and continue firing until directed otherwise.

- Men not in the killing zone maneuver, as directed, against the ambush force.

- The attack is continued, as directed, to eliminate the ambush or to break contact.

In each situation, the success of the counterambush drill employed is dependent on the patrol members being well trained in recognizing the nature of an ambush and well rehearsed in the proper reaction.

USE OF IMMEDIATE ACTION DRILLS

In chance contact and in ambush, the immediate action drills that a patrol uses and the patrol's subsequent actions are determined largely by whether the assigned mission prohibits or permits contact (except that necessary to accomplish the mission). The immediate action drill *freeze*, however, can be used in either circumstance and not affect the assigned mission (Figure B-1).

Some immediate action drills may be used repeatedly with little danger that frequent use will enable the enemy to develop effective countermeasures. *Freeze* and *hasty ambush* are in this category. The situations in which their use is appropriate do not lend themselves to easy conversion by the enemy into a baited trap.

Habitual use of some immediate action drills can be very dangerous, however. For example, too frequent use of *immediate assault* can lead the enemy to expose a small force to an apparently undetected patrol, causing the patrol to launch an *immediate assault* into the massed fires of a larger, concealed force. This countermeasure has been very effectively used by the Vietcong guerrillas in South Vietnam.

Any immediate action drill must be carefully studied to detect any potential dangers that may arise from frequent use. If these dangers cannot be eliminated, the drills must be varied to avoid setting patterns.

B-3

189

MISSION REQUIREMENT	SITUATION	IMMEDIATE ACTION DRILL	SUBSEQUENT ACTION
Avoid contact, if possible, and quickly break any contact made.	The patrol sees the enemy approaching. The possibility appears good that the enemy will not detect the patrol. The enemy is so close that there is no time to establish a hasty ambush. 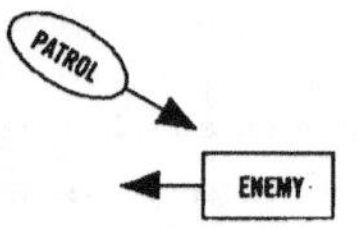	FREEZE. Every man freezes in place, weapons at the ready. The patrol opens fire only if detected. 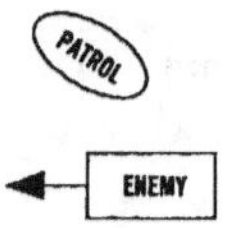	If no contact is made, the patrol continues after the enemy passes ①. If contact is made, the patrol moves quickly into line formation ②, assaults immediately ③, breaks contact with the enemy ④, and continues; or breaks contact immediately after initial fire ⑤ by using the *clock system* ⑥, and the patrol continues. 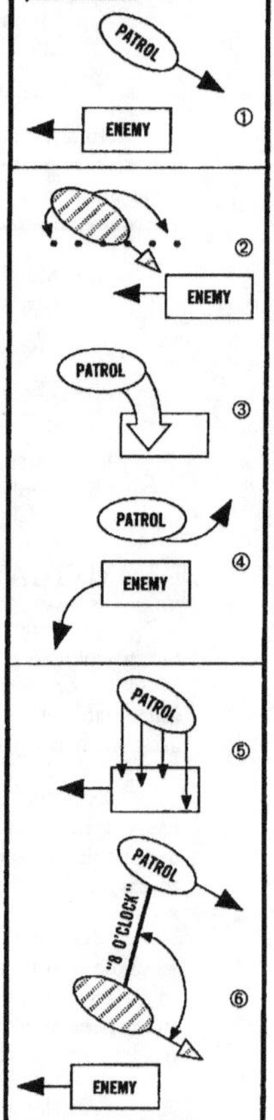

Figure B-1. Immediate action drills.

B-4

190

MISSION REQUIREMENT	SITUATION	IMMEDIATE ACTION DRILL	SUBSEQUENT ACTION
Avoid contact, if possible, and quickly break any contact made (continued).	The patrol sees the enemy approaching. Contact appears unavoidable.	FREEZE, followed by HASTY AMBUSH. The patrol executes ambush only if detected.	If no contact is made, the patrol continues after the enemy passes ①. If the ambush is executed, the patrol assaults by fire only ②, withdraws quickly ③, and continues.
	The patrol and the enemy detect each other at the same time and at such close range that breaking by the *clock system* or fire and maneuver is not appropriate.	IMMEDIATE ASSAULT ① to the enemy if it withdraws ②; through the enemy if it stands fast ③.	Break contact and continue mission.

Figure B-1. Immediate action drills (continued).

MISSION REQUIREMENT	SITUATION	IMMEDIATE ACTION DRILL	SUBSEQUENT ACTION
Avoid contact, if possible, and quickly break any contact made (continued).	The patrol is ambushed (*near* ambush). 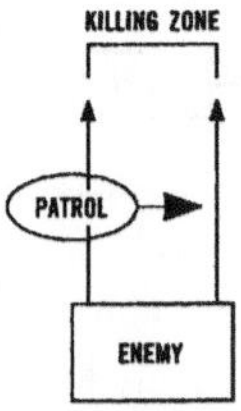	COUNTERAMBUSH. Men in the killing zone assault attack force. Other men attack, as directed, to permit entire patrol to break contact. 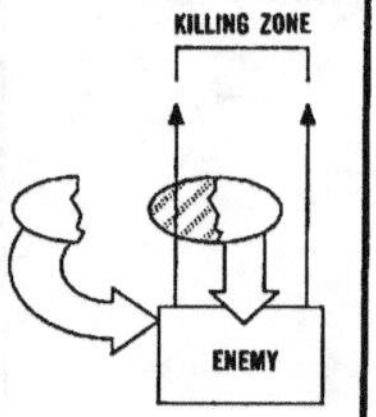	Reorganize and continue mission. 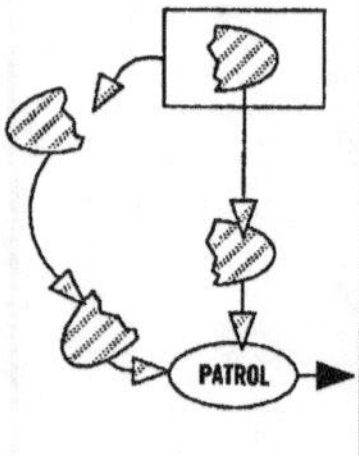
	The patrol is ambushed (*far* ambush). 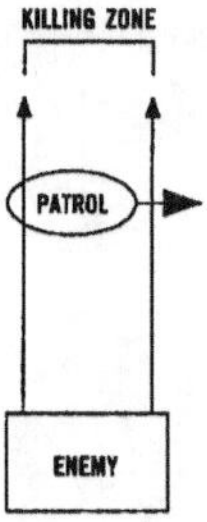	COUNTERAMBUSH. Men in the killing zone return fire, seek cover, continue firing. Other men attack, as directed, to permit the patrol to break contact. 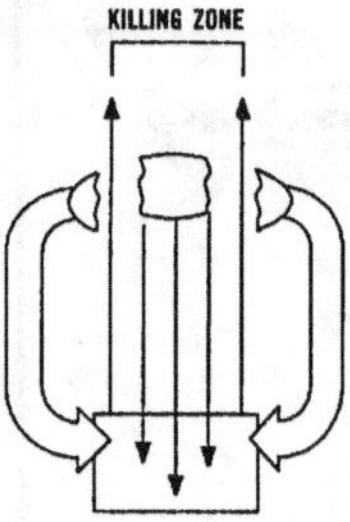	Reorganize and continue mission. 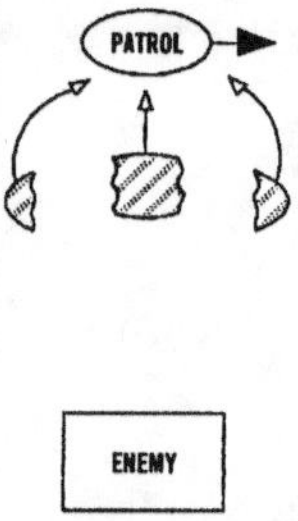

Figure B-1. Immediate action drills (continued).

MISSION REQUIREMENT	SITUATION	IMMEDIATE ACTION DRILL	SUBSEQUENT ACTION
Engage targets of opportunity and otherwise exploit opportunities to engage in decisive combat.	The patrol sees the enemy approaching. The possibility appears good that the enemy will not detect the patrol. The enemy is so close that there is no time to establish a hasty ambush.	FREEZE.	The patrol allows the enemy to advance as close as possible ①. The patrol leader opens fire when the enemy is in the most vulnerable position (any patrol member opens fire if detected) ②. The patrol assaults the enemy with great violence and heavy fire, destroys it, or pursues it if appropriate ③.

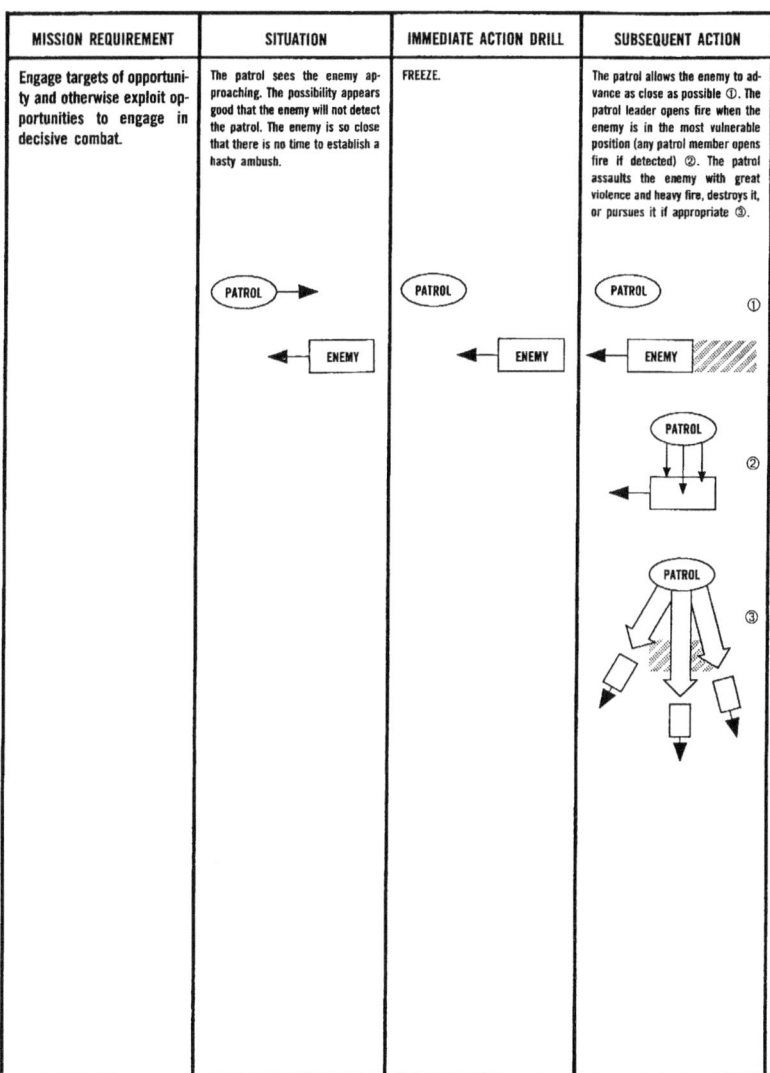

Figure B-1. Immediate action drills (continued).

MISSION REQUIREMENT	SITUATION	IMMEDIATE ACTION DRILL	SUBSEQUENT ACTION
Engage targets of opportunity and otherwise exploit opportunities to engage in decisive combat (continued).	The patrol sees the enemy approaching. Contact appears unavoidable. 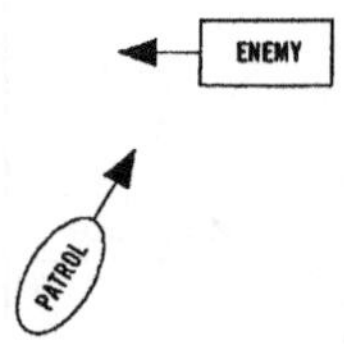	FREEZE, followed by HASTY AMBUSH. 	The patrol executes the ambush when the enemy is in its most vulnerable position, assaults the enemy with great violence and heavy fire, destroys it, or pursues it if appropriate. 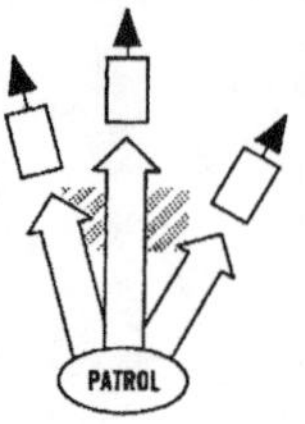
	The patrol and the enemy detect each other at the same time. 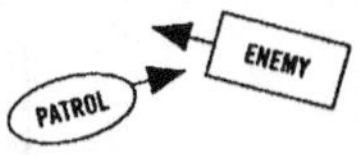	IMMEDIATE ASSAULT. 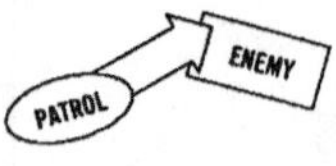	The patrol continues the assault until the enemy is destroyed. Escaping enemy is pursued and destroyed. 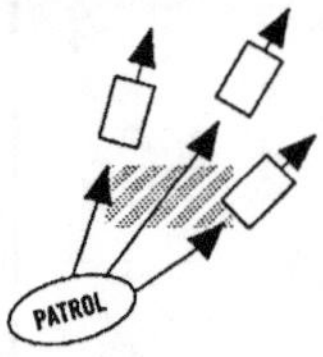

Figure B-1. Immediate action drills (continued).

MISSION REQUIREMENT	SITUATION	IMMEDIATE ACTION DRILL	SUBSEQUENT ACTION
Engage targets of opportunity and otherwise exploit opportunities to engage in decisive combat (continued).	The patrol is ambushed (*near* ambush).	COUNTERAMBUSH. Men in the killing zone assault to destroy the attack force. Other men attack, as directed, to eliminate ambush.	The escaping enemy is pursued and destroyed.
	The patrol is ambushed (*far* ambush).	COUNTERAMBUSH. Men in the killing zone return fire, seek cover, and continue firing. Other men attack, as directed, to enable men in the killing zone to maneuver ①. All attack, as directed, to eliminate ambush ②.	Escaping enemy is pursued and destroyed.

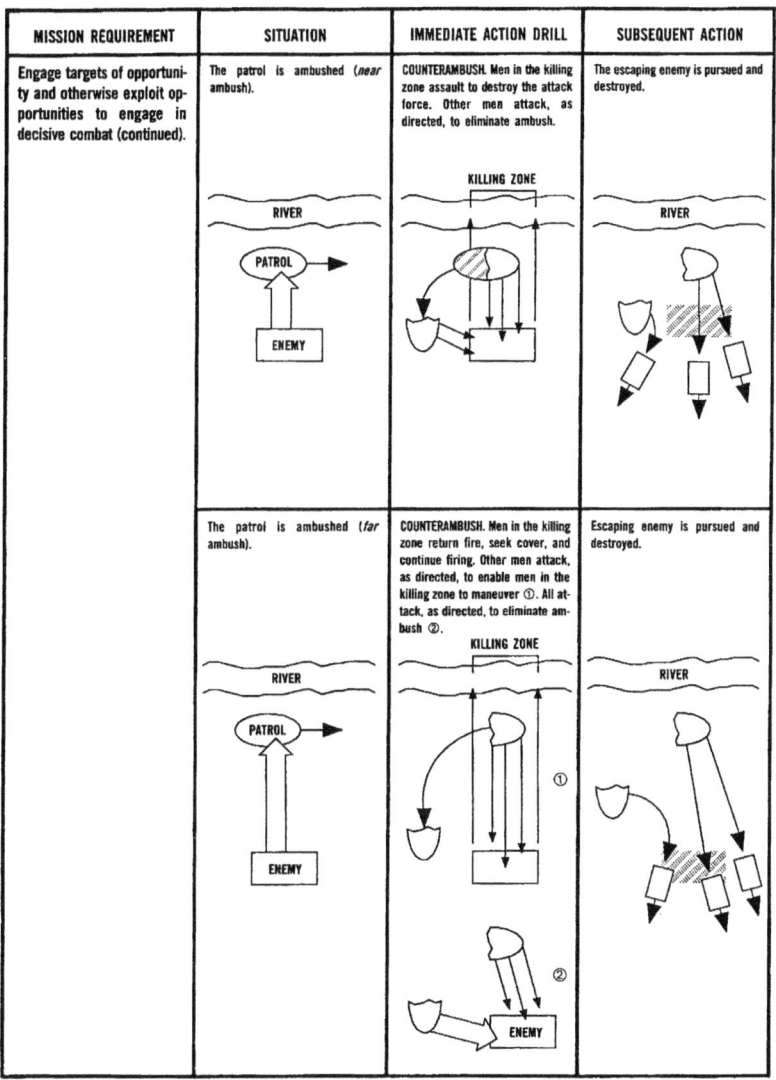

Figure B-1. Immediate action drills (continued).

MISSION REQUIREMENT	SITUATION	IMMEDIATE ACTION DRILL	SUBSEQUENT ACTION
The immediate action drill used is not affected by the assigned mission.	An aircraft (enemy or unidentified) that may observe the patrol is sighted or heard. 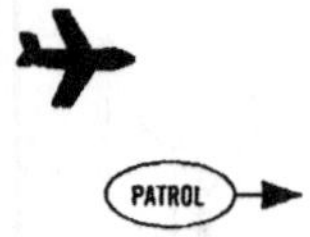	FREEZE. Initiated by the first man sighting or hearing aircraft. 	As directed by patrol leader. 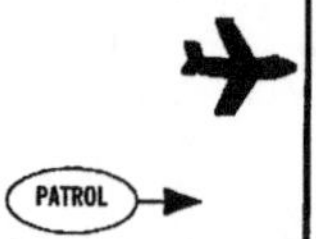
	Enemy aircraft makes low-level attack. 	AIRCRAFT ATTACK. Initiated by the first man detecting aircraft approach. Men seek cover, placing themselves perpendicular to the direction of attack (to present a shallow target). 	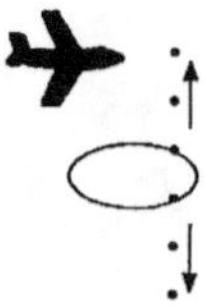 If aircraft does not repeat attack, continue mission ①. If aircraft returns or if more than one aircraft attacks, continue dispersion and seek more cover ②. 

Figure B-1. Immediate action drills (continued).

RECONNAISSANCE PATROL

Section I. General Tips

While on a mission, minimize fatigue; tired men become careless.

If you show confidence, your men will have confidence.

If the patrol leader loses his temper, it will effect his judgment. Keep cool and think ahead; always keep an alternate plan in mind. Don't be afraid to take advice from your men.

Team work, the key to success, only comes through constant practice and training. Realism must be injected into all phases of training, such as zeroing of weapons at targets in the jungle, use of live training aids for prisoner-of-war (PW) snatch or ambush practice.

Men who have a good physical training program have fewer health problems.

Make sure that personnel take salt tablets as a preventive measure rather than waiting until collapse is imminent. One tablet in a canteen of water is a good way to take salt, especially on very hot and humid days.

If your mission calls for emplacing a mine in a road, take an extra fuse along in case one is lost.

All personnel should wear loose fitting and untailored clothing on field operations. Tight fitting clothing often tear or rip, allowing easy access to exposed parts of the body for mosquitoes and leeches.

Each patrol leader should have a pre-mission and postmission checklist to ensure that nothing is left behind.

Use tact when reprimanding personnel, especially indigenous patrol members. If possible, take the man aside to criticize him. This enables him to reason positively to the criticism and to not feel ridiculed and lose self-confidence.

Do not hang clothing or bandannas on green bamboo if you plan on wearing them afterward. The fuzz on the bamboo is just like itching powder.

Conduct English classes for indigenous personnel, especially interpreters. Conduct classes for US personnel on the indigenous patrol members' dialect.

Preset frequencies on the radio so that a quick turn of the dial will put you on the desired frequency. This is especially helpful at night when you want to avoid using a light.

Carry CS powder in plastic insect repellent or lube oil bottles. It is difficult to put CS powder in them but it is definitely worth the effort. Sprinkle CS powder in and on empty C

C-1

ration cans and food containers. This will prevent animals from digging them up once you have buried them.

Section II. Weapons Tips

Tape the muzzle of your weapon to keep out water and dirt. Leave lower portion of slits open for ventilation.

Use a full magazine of tracer during infiltration and exfiltration. If taken under fire during infiltration or exfiltration, the tracers can be used to identify enemy positions to friendly air assets.

The last three rounds in each magazine should be tracer. This reminds the firer that he needs a loaded magazine.

Replace the cartridge in the chamber of your weapon each morning *quietly*. Condensation may cause a malfunction.

Oil the selector switch on your weapon daily and work the switch back and forth, especially during the rainy season. This prevents the switch from sticking—a common occurrence.

Always carry a small vial or tube of lubricating oil for your weapon.

Always carry your weapon with the selector switch on *safe*.

To improve noise discipline, tape all sling swivels or remove them from weapons.

During extraction do not fire weapons from helicopters after leaving the landing zone (LZ) because a gunship may be passing under you without your knowledge.

Do not retrieve your first expended magazine during contact because it will consume valuable time.

Check all magazines before going on an operation to ensure they are clean and properly loaded.

Never assume that your weapon is clean enough on an operation. CLEAN YOUR WEAPON DAILY.

Place magazines upside down with bullets pointed away from your body in the ammunition pouches. This keeps dirt and water out of them and prevents injury to you by your own ammunition if rounds go off due to enemy fire.

Section III. Reconnaissance Patrol Tips

When making a visual reconnaissance (VR), always mark every LZ within the AO and near it on your map. Plan the route of march so that you will always know how far and on what azimuth the nearest LZ is located.

Do not cut off too much of the map showing your reconnaissance zone (RZ). Always designate at least 5 to 10 kilometers surrounding your RZ as running room.

Base the number of canteens per mission upon the weather and availability of water in the AO. Select water points when planning your route of march.

Check all patrol members' pockets prior to departing home base for passes, identification cards, lighters with insignias, and rings with insignias. Personnel should only carry dog tags while on patrol.

If the patrol uses a grenadier armed with rifle grenades, have him place a crimped cartridge as the first round in each magazine carried. After firing the grenade, he can use the rifle normally. When the magazine is empty and a new one inserted the grenadier can then quickly fire another grenade.

Always carry maps and notebooks in waterproof containers.

Use a pencil to make notes during an operation. Ink smears when it becomes wet, whereas lead does not.

Inspect each patrol member's uniform and equipment, especially radios and strobe lights, prior to departure on a mission.

If you use the Hanson Rig, adjust your harness and webbing before leaving on patrol.

During the rainy season take extra cough medicine and codeine on patrol.

All patrol members should know how to properly administer morphine, who in the patrol is carrying it, and where it is on that person(s).

All survival equipment should be tied or secured to the uniform or harness to prevent loss if pockets become torn.

Each US or key patrol member should carry maps, a notebook, and communications-electronic operation instruction (CEOI) in the same pocket of each uniform for hasty removal by other members if one member becomes a casualty.

Take paper matches or disposable cigarette lighters to the field in a waterproof container. Do not take zippo-type cigarette lighters as they make too much noise when opening and closing.

Tie panel and mirror to pocket flap to prevent losing.

Always carry rifle cleaning equipment on an operation, for example, a brush, oil and at least one cleaning rod.

Each patrol should have designated primary and alternate rallying points at all times. The patrol leader is responsible for ensuring that each patrol member knows the azimuth and approximate distance to each rallying point and LZ.

Never take pictures of patrol members while on patrol. If the enemy captures the camera, they will have gained invaluable intelligence.

At least two penlights should be taken by each patrol.

While on patrol, move 20 minutes and halt and listen for 10 minutes. Listen half the amount of time you move. Move and halt at irregular intervals.

Stay alert at all times. You are never 100 percent safe until you are back home.

C-3

Never break limbs or branches on trees, bushes, or palms, for this leaves a very clear trail for the enemy to follow.

Put insect repellent around tops of boots, on pants fly, belt, and cuffs to repel leeches and insects.

Do most of your moving during the morning hours to conserve water. However, never be afraid to move at night, especially if you think your remain–overnight (RON) location has been discovered.

Continually check your point man to ensure that he is on the correct azimuth. Do not run a compass course on patrol, change direction regularly.

If followed by trackers, change direction of movement often and attempt to evade or ambush your trackers; they make good PWs.

Do not ask for a *fix* from the forward air controller (FAC) unless absolutely necessary. This will aid in the prevention of compromise.

Force yourself to cough whenever a high–performance aircraft passes over. Cough will clear your throat, ease tension, and cannot be heard. When there is no noise and you must cough, cough in your hat or neckerchief to smother the noise.

Never take your web gear off, day or night, except when changing clothes. For example, in an area where it is necessary to put the jungle sweater on at night, no more than two patrol members at a time should do so. Take the sweaters off the next morning to prevent cold and overheating.

If you change socks, especially in the rainy season, try to wait until you reach the RON site and have no more than two patrol members change socks at one time. Never take off both boots at the same time.

When a patrol member starts to come down with immersion foot, stop in a secure position, remove the injured persons boots, dry his feet, put foot powder on them and place a ground sheet or poncho over his feet so that they can dry out. Continued walking will make matters worse and cause the man to become a casualty, thereby halting further progression of the patrol.

Desenex or Vaseline rubbed on the feet during the rainy season or in wet weather helps prevent immersion foot. It will also help avoid chapping if put on the hands.

Gloves protect hands from thorns and aid in holding a weapon when the weapon heats up from firing.

Place a plastic cover on your radio to keep it dry in the rainy season.

When using a wiretap device, never place the batteries in the set until needed. If the batteries are carried in the device they will lose power even though the switch is in the off position.

If batteries go dead or weak do not throw them away while on patrol. Small batteries can be recharged by placing them in arm pits or between the legs of the body. A larger battery can gain added life by sleeping with the battery next to the body. Additional life can also be gained by placing batteries in the sun.

If possible, carry an extra hand set for the radio and ensure that it is wrapped in a waterproof container.

C-4

Always carry a spare radio battery, but do not remove the spare from its plastic container prior to use or it may lose power.

Do not send *same* or *no change* when reporting patrol location. Always send your coordinates. Keep radio traffic at a minimum.

Avoid over confidence; it leads to carelessness. Just because you have seen no sign of the enemy for 3 or 4 days does not mean that it is not there or has not seen you.

A large percentage of patrols have been compromised due to poor noise discipline.

Correct all patrol individual errors as they occur.

All personnel should camouflage faces and back of hands in the morning, at noon, and at RON or ambush positions.

Never cook or build heating fires on patrol. No more than two persons should eat chow at any one time. The rest of the patrol should be on security.

When the patrol stops, always check out area 40 to 60 meters from the perimeter.

All patrol members should take notes while on an operation and compare them nightly. Each man should keep a list of tips and lessons learned and add to them after each operation.

Each man in a patrol must continually observe the man in front of him and the man behind him, in addition to watching for other patrol members' arm and hand signals.

A reconnaissance patrol should never place more than one mine, anti- personnel (AP) or antitank (AT), in one small section of a road or trail at a time. If more than one is set out, the patrol is just resupplying the enemy, because when a mine goes off, a search will be made of the immediate area for others, and they will surely be found.

During the dry season, do not urinate on rocks or leaves, but urinate in a hole or small crevice. The wet spot may be seen, and the odor will carry further.

When crossing streams, observe first for activity, then send a point man across to check the area. Then cross the rest of the patrol members, with each taking water as he crosses. If in a danger area, have all personnel cross prior to getting water. Treat all trails (old and new), streams, and open areas as danger areas.

Carry one extra pair of socks, plus foot powder, on a patrol, especially during the rainy season. In addition, each patrol member should carry a large sized pair of socks to place over his boots when walking or crossing a trail or stream.

During rest halts do not take your pack off or leave your weapon alone. During long breaks, such as noon chow, do not take your pack off until your perimeter has been checked for at least 40 to 60 meters out for 360 degrees. During breaks throw nothing on the ground. Either put the trash in your pocket or spray it with CS powder and bury it.

In most areas, the enemy will send patrols along roads and major trails between the hours of 0700 and 1000 and from 1500 to 1900. Since most of the enemy's vehicular movement is at night, a patrol that has a road watch mission should stay no less than 200 meters from the road during the day and move up to the road just prior to last light. When the enemy makes a security sweep along a road, usually twice a week, he normally does not check further than 200 meters to each flank.

C-5

If you hear people speaking, move close enough to hear what they are saying. The reason is obvious. The patrol leader should make notes.

While on a patrol, do not take the obvious course of action and do not set a pattern in your activities, such as, always turning to the left when *button hooking* to ambush your own back trail.

A dead enemy's soldier's shirt and the contents in his pockets, plus his pack (if he has one), are normally more valuable than his weapon.

If the enemy is pursuing you, you should deploy delay grenades or delay Claymores of 60 to 120 seconds. In addition, throw CS grenades to your rear and flanks. Give the enemy a reason or excuse to quit following you.

Do not fire weapons or use Claymores or grenades if the enemy is searching for you at night. Use CS grenades instead. Using CS grenades will cause the enemy to panic and will not give your position away. You can move out in relative safety while they may end up shooting each other. If Claymores become necessary, use time–delayed or time–delayed white phosphorous (WP) Claymores.

Section IV. Remain Overnight Tips

Practice proper RON procedures when your patrol is training, even if you are on a rifle range. Take advantage of all training opportunities. Many training areas are not in what we could call *safe zones*.

Select a tentative RON site (from you map) at least 2 hours in advance.

Deviate from your route of march often. Never move in a straight line.

After passing a suitable RON site *fishhook* and move into your selected position so that you can observe your own trail.

When in position, personnel should keep their equipment on and remain alert until the perimeter has been checked for 360 degrees at a distance of no less than 40 to 60 meters.

Packs should not be taken off until it is dark.

Before dark each patrol member should memorize the azimuth and distance to the trees and bushes around his RON site.

When deploying the patrol to RON, place the point man in a position opposite the most likely avenue of approach to lead the patrol out in case of emergency.

If a patrol is within range of friendly artillery and has preplanned concentrations, azimuths should be taken (observer target (OT) line) noting distances prior to night fall. Nearby large trees or pre–positioned stakes will aid as hasty reference points for calling in artillery at night.

If it is necessary to send in a nightly situation report (SITREP), do not send the message from your RON position. Send your present location but add that you will RON 100 meters east or 200 meters north. This will confuse the enemy as to your exact location if he has monitored your transmission with direction finding equipment. Use your CEOI.

C–6

Keep transmissions to a minimum. It is better to send the location of your RON position the next morning, after you have moved out. The enemy may monitor your traffic but he will not know in what direction you plan to move.

Do not send radio transmissions from your RON site unless they are necessary. Be prepared to move if you do send radio transmissions.

Prior to dark, the patrol leader should tell each man the primary and alternate rallying points.

Half of the patrol members should have their compasses set on the primary rallying point and the other half on the alternate. If the enemy comes from the direction of the primary rallying point, the man with the azimuth of the alternate rallying point on his compass can lead the patrol out.

A buddy system should be established in case casualties are taken at night. Each man will take care of another man and his equipment if one is wounded, injured, or killed.

The pack or rucksack can be used as a pillow, however, ensure that the carrying straps are in the *up* position for easy insertion of the arms in case of rapid withdrawal.

It is permissible to unhook the web gear or harness, but it should not be taken completely off at night or at any other time during the entire stay in the field.

If a person coughs or talks in his sleep, make him sleep with a gag in his mouth.

United States patrol members should not *bunch up* or sleep next to each other. One grenade or automatic burst from a weapon could get them all. Each patrol member should be able to communicate with each other without moving from position using tug lines or hand and arm signals.

Check the overhead vegatation of your RON position to ensure there is an opening for the strobe light to direct air support in the event of enemy contact.

Know what the next day's plans are before settling down for the night.

When placing Claymores around your RON site, they should be placed one at a time by two men, one man emplacing the mine while the other stands guard. Never emplace Claymores in a position that prevents you from having visual contact with it.

Claymores should be emplaced so that the blast parallels the patrol, ensuring that the firing wire does not lead straight back to the patrol's position from the mine. If the Claymores are turned around by the enemy they will not point at the patrol.

Determine, in advance, who will fire each Claymore and who will give the command or signal.

In most instances it is better not to put out Claymores around RON positions, but to rely on the use of CS grenades for the following reasons:

- When Claymores have been put out, and the enemy is discovered moving in on the patrol, the patrol has a tendency to stay in place too long, allowing the enemy to get within the killing zone.

- If the patrol discovers the enemy moving in on its, the enemy will normally be *on line*, not knowing the exact position of the patrol. If no Claymores are out,

C-7

predesignated patrol members throw CS grenades in the direction of the enemy. After the gas begins to disperse, the patrol can withdraw. When an enemy soldier is hit with the CS he will normally panic. If he has a gas mask with him, and puts it on, he can no longer see clearly. If he does not have one, he will run away and may even fire his weapon indiscriminately, causing overall confusion and panic. In either case, the patrol has a good chance to escape—unharmed and unseen.

- If a Claymore is triggered, a grenade thrown, or a rifle fired, the enemy may flank the patrol and box it in.

All patrol members should be awake, alert, and ready to move prior to first light.

Another check of the perimeter (360 degrees), at a distance of at least 40 to 60 meters, should be made prior to moving out or prior to retrieving Claymores.

A thorough check should be made of the RON site to ensure that nothing is left behind and that the entire site is sterile.

The patrol leader must make sure that each man takes his daily malaria tablet.

Never eat chow or smoke cigarettes in the RON position. The odor of the food or tobacco gives the position away.

Be alert when leaving the RON site. If you have been seen, you will probably be attacked or ambushed within 300 meters.

Patrol leaders should check themselves to ensure they do not form the common habit of constantly turning to the left, or right, when *fish-hooking*.

Habits are easily formed, as mentioned previously around certain times of the day. For example, some always move into a RON site at 1830 or into a noon break position at exactly 1000 each day. If the enemy has been observing you, it will take note of this and will plan an ambush for you.

Section V. Breaking Out of Encirclement Tips

GENERAL

Too many times reconnaissance patrols that have not planned for or practiced methods to *break out* from an encirclement have been encircled by the enemy. The following methods and suggestions have worked for others in the past and hopefully will help you if you find your patrol in such a situation.

If the enemy encircles the patrol, the sooner you attempt to break out the better chance you will have to do so effectively and with the least amount of casualties. The longer you wait the stronger the enemy becomes.

PREPARATIONS FOR BREAKING OUT OF ENCIRCLEMENT

Plans must be made prior to break out to take care of the following:

- Rucksacks and equipment left behind must be destroyed by someone.
- The dead must be left behind. Someone must remove any classified documents, such as CEOI, notebooks, maps.

C-8

- During the breakout assault, one or two persons, depending upon the size of the patrol, provide rear security. This will include assisting any personnel who may be wounded before or during the attempt. Additionally, this person(s) should recover documents from personnel killed during the breakout movement. No attempt should be made to take the dead with the patrol.

- Keep in mind that the successful completion of the mission depends on getting the information back to headquarters. All personnel must be reminded of the important information the patrol has observed.

FORMATION TO USE

The most effective method a small element can use (5– to 12–man patrols) is to form into a pyramid configuration, with the base of the pyramid leading. The following actions should take place:

- The patrol forms into position.
- CS rounds from M203s or CS grenades are fired or thrown to the flanks.
- White phosphorous grenades are thrown to the rear.
- A Claymore mine or fragmentation grenades are fired or thrown in the direction the patrol will move.
- Immediately after the Claymore or grenades go off to the front, the patrol moves out.
- The first element of the line will fire weapons on full automatic. The others hold fire.
- When the first element's magazines are empty, the second element moves through it and continues to fire.
- When the second element has emptied its magazines, the first element will have reloaded and will pass through the second element, taking up the assualt but will only fire on semiautomatic.
- Once the patrol starts to move it must move rapidly, but not run, and never stop until completely out of the encirclement.

SUPPORTING FIRES

Artillery, helicopter gunships, and tactical air (TACAIR) fire support, if available, should be used to assist breakout. These are discussed below.

Supporting artillery fire, within range, can be effectively employed to pave the way out of an encirclement or near encirclement. When foul or inclement weather prevents the use of helicopter or TACAIR for support, you must use artillery if available. Artillery support, when available, should also be requested at the first sign of trouble, for many times it can be firing in your direction before air support arrives on station to assist you. As you learned in basic training, when you are subjected to incoming mortar or artillery fire, move out of the area as quickly as possible. Enemy forces follow this same doctrine. When you desire to break out with the aid of artillery, first have the fires placed completely around your position. Then, having selected your desired heading, *walk* the artillery in front of you. This technique will effectively lead you out of the danger area and you may even pick up a shell–shocked or wounded PW on your way out.

Helicopter gunships can assist you with almost continuous close-in fire support, firing directly in front and to the rear during the breakout. The effect this fire will have depends on

C–9

the density of the vegetation, location of the patrol, and whether or not the supporting aircraft crews can see you or your signals. You may have to direct their fire by adjusting from the strike of the rounds and rockets.

Tactical air strikes can assist in break out of an encirclement. To do this, call for bombs in the direction you desire to move. Since the enemy will get as close to the patrol as possible to avoid air strikes, you should call in the bombs first and then have the TACAIR fire its machine guns in front of you as you move out. Machine-gun fire can be placed much closer to you than bombs.

Section VI. Movement Technique Tips

MOVEMENT TECHNIQUES

There are five basic techniques of movement that can be employed by small reconnaissance patrols to avoid being detected or encircled by the enemy. Each of these are explained and discussed below.

Box Technique

This is a simple and effective method to use and it takes very little practice to employ (Figure C-1). From a given point, the patrol moves out on a set azimuth for a specifically set number of meters or paces, for example 35 meters. The patrol then makes a 90-degree turn and moves 75 meters, then another 90-degree turn for 30 meters, another 90 degree turn for 30 meters, and another for 30 meters. You will have formed a *box*. At this point you may do any one of several things. You may wait in ambush for your trackers or pursuers; walk backward across your old trail, if the vegetation and soil is such that it is impossible to hide your tracks; or you may continue on. When you move out, after having formed the first box, move for another 50 to 75 meters and form another box. By forming these boxes, it will enable the patrol to ambush the pursuers and will definitely confuse any trackers as to your direction of movement. It will also discourage the enemy if you occasionally booby-trap your back trail. You can maintain a general compass heading without the enemy becoming aware of your presence until you are out of the danger area or until it loses you completely. A word of caution though: Do not continually make the boxes the same size or continually turn to the right or left. Never set a definite pattern of movement.

Figure Eight Technique

The *figure eight* method (Figure C-2) is very similar to the box technique in that you do basically the same thing, except here you make circles instead of squares.

Angle Technique

The *angle* technique (Figure C-3) is another effective method to use in evasion and takes very little practice to employ. The patrol changes its direction of movement from its present patrol route of march in a series of angle movements. For example, the patrol will make an angle move to change direction such as 30 degrees, 45 degrees, 70 degrees for a hundred or so meters, then do it again to confuse the enemy.

C-10

206

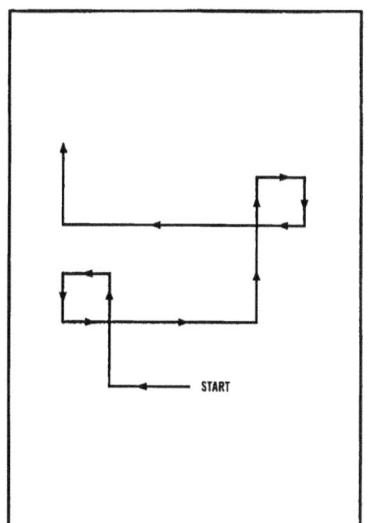

Figure C-1. Box technique.

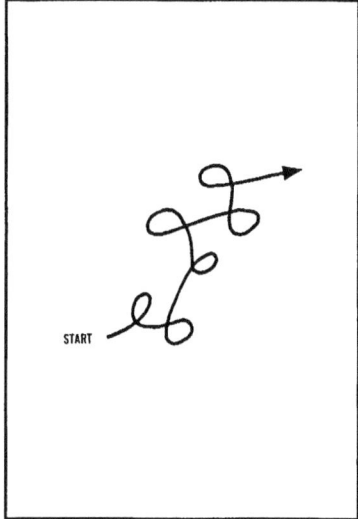

Figure C-2. Figure Eight technique.

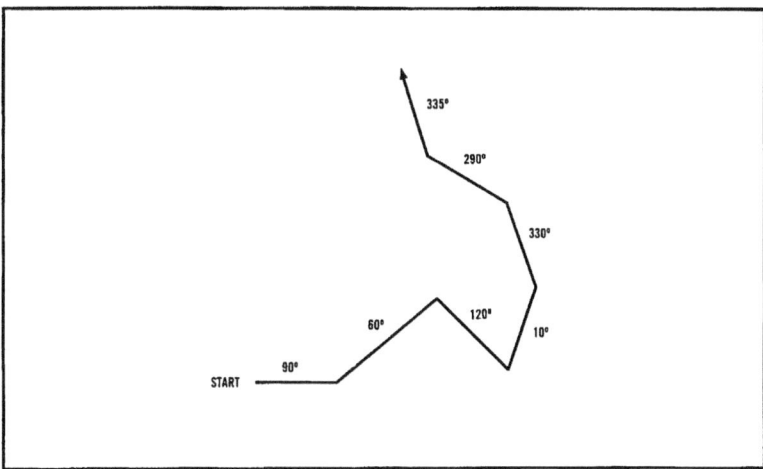

Figure C-3. Angle technique.

C-11

207

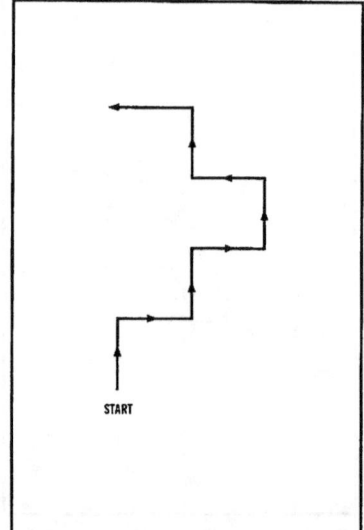

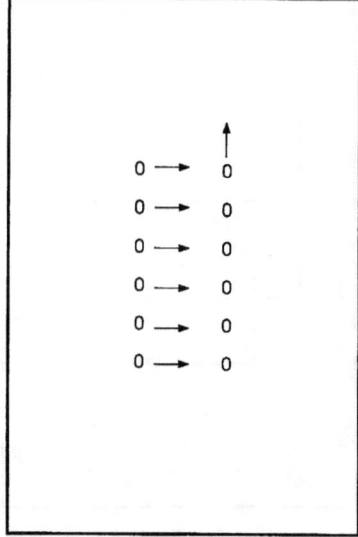

Figure C-4. Step technique.   **Figure C-5. Skip technique.**

Step Technique

The *step* technique (Figure C-4) is a simple method of changing the route of march in 90-degree turns for a distance of a hundred or so meters.

Skip Technique

The *skip* technique (Figure C-5) is an effective method that requires practice to employ. The patrol stops in place and, on command, moves left or right of the present route. Each member moves as carefully as possible to prevent making a trail or leaving telltale signs to the flank for a distance of 20 to 30 meters and then resumes the patrol's former route of march. The patrol leader should send the point man ahead to make a false trail for 30 to 50 meters before using the skip method. This method takes practice and patrol members have to be careful not to leave signs as they move.

ADDITIONAL INFORMATION

Never set a pattern. If one technique does not work, change to another.

In both the box and figure eight techniques, the size of the squares or circles will depend on the terrain and vegetation. The box technique is extremely effective at night. Both methods can be used to find a hole or weak point in the enemy's encircling perimeter from which to break out. Both techniques have been used successfully in the past by reconnaissance patrols. One survivor of an ambushed reconnaissance patrol succeeded for 3

C-12

days in ambushing and killing six enemy pursuers. He employed the figure eight method before being spotted and recovered by searching aircraft.

During the dry season, CS powder spread over your back trail is extremely helpful in stopping dogs.

During the rainy season, CS powder is almost useless or very ineffective against dogs. It is much more effective to drop a CS grenade during wet weather since it hangs low to the ground and remains effective against enemy personnel, especially those who do not have or carry protective masks.

Your tactics and techniques are only as good or effective as you make them. Success is achieved only through constant practice, training, and rehearsals.

APPENDIX D

TWELVE–POINT CACHE REPORT FORMAT

TYPE OF CACHE

The component for which the cache is intended (guerrilla unit, sabotage cell, operator), and the functional purpose of the items (weapons, demolitions, communications) to be cached.

METHOD OF CACHING

Burial, concealment, or submersion.

CONTENTS

An itemized list of all items in each container and a description of how each item is packaged.

DESCRIPTION OF CONTAINERS

The size, weight, and other descriptive details. If several containers are included in the cache, assign a number to each container. Ensure these numbers appear on the sketch of the cache so that each container is identified by its position in the cache.

GENERAL AREA

The generally recognizable place names. Ordinarily, place names include the country, province, and smaller political divisions, down to the nearest town or village.

IMMEDIATE AREA

The initial reference point and instructions for proceeding from this point to the FRP. All landmarks that facilitate visual recognition of the route should be described.

CACHE LOCATION

The FRP and the exact sightings, linear measurements for pinpointing the cache. All measurements must be stated in linear units (meters, feet) that recovery personnel can understand and use.

EMPLACEMENT DETAILS

All features of the site or natural conditions that must be considered for physically retrieving the cache. The following represent the essentials, depending upon the method of caching.

Burial

The exact depth underground of each container, precise description of shoring (if used), all known seasonal variations (surface vegetation, date, and depth of ground freezing and thawing). The type of soil and the time required for emplacement also provide useful guides for planning the recovery operation.

Concealment

Exactly how the cache is placed in the site and any physical covering (plaster, bricks) that must be penetrated or removed to recover the cache. Provide full instructions if removing or replacing the covering involves any special problems or techniques (matching the plaster or mortar). All necessary information about a custodian, if one is used, should be included.

Submersion

Depth of the water (including high- and low-water marks), submersion depth (if the container does not rest on the bottom of the lake or river), type of bottom water motion, clearness of the water, and usual freezing and thawing dates.

OPERATIONAL DATA AND REMARKS

List of equipment needed for recovery of the cache. Description of at least two routes to the site that offer maximum natural concealment and a means of escape in case of sudden attack. Nearby houses and thoroughfares; description of local security forces, their regular posts, and patrol routes in the vicinity of the cache. Suggestions for action cover when visiting the site, including warning of what cover to avoid, and any other information that may ease planning the recovery operation. Give special consideration to any equipment that may be needed for recovery, even though it was not used in emplacement.

DATES OF EMPLACEMENT AND DURATION OF THE CACHE

Based on an estimate of how long the contents of the cache will remain usable. Pertinent factors include the normal shelf life of items that deteriorate with time (medicine, batteries); the expiration date of official documents (passports, licenses); how long the packaging will withstand moisture, penetration, and corrosion.

SKETCHES AND DIAGRAMS

Whatever sketches and diagrams are necessary to illustrate the instructions for locating the cache and the description of the cache. Include at least an areal sketch, showing the route from the initial reference point to the FRP (Figure D-1), and a site diagram, showing precisely how the cache is pinpointed (Figure D-2). Photographs of the immediate area, the initial reference point, the FRP, and other landmarks in the vicinity of the site are not essential, but they may be helpful.

RADIO MESSAGE FOR RECOVERY

It is useful to have such a message drafted in case an emergency dictates its use; though a radio message may never be required for recovery. The best time for drafting the message is when the details are fresh in the mind of the emplacer. The radio message should include type of cache, method of caching, and concise instructions for locating the site. The acid test of these instructions is whether they can be reduced to a message that is unmistakably clear, but brief enough for secure radio transmission. Give very careful consideration to the intended recovery person's familiarity with the area, as well as what maps and makeshift surveying instruments will be available to him. The message must be in a language he is sure to understand. It must be drafted or translated by someone who is fluent in the language.

D-2

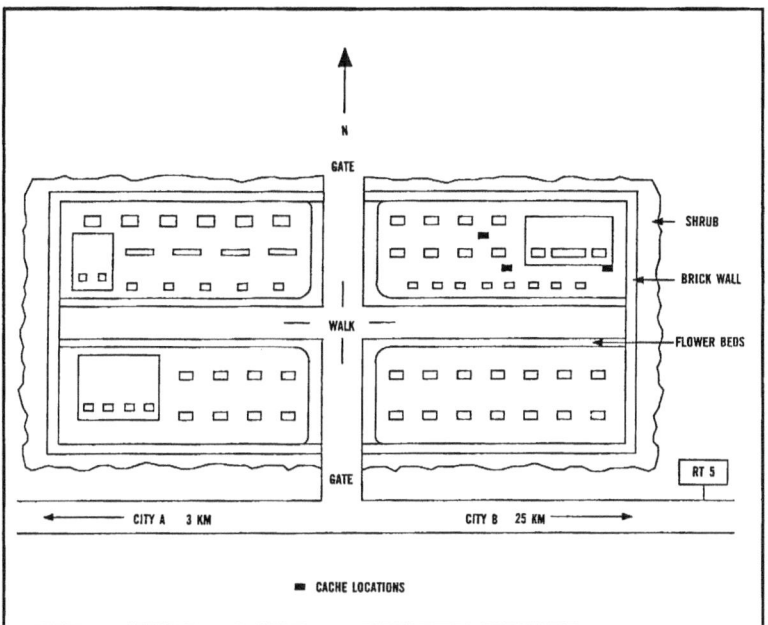

Figure D-1. Cache locations.

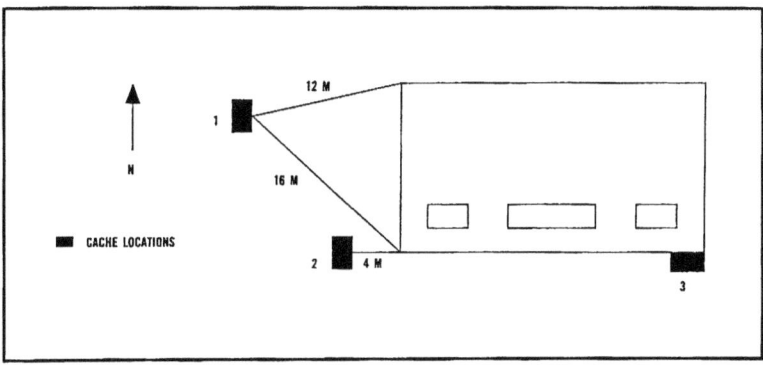

Figure D-2. Cache locations.

The radio message in Figure D-3 gives instructions for recovering the three-point cache illustrated in sketches D-1 and D-2.

This sample message is intended to illustrate the absolute minimum data that ordinarily is essential for recovery. Additional data should be included in a radio message only when special circumstances require it. For instance, if a cache is too heavy or too large for one man to carry, the weight or the exterior dimensions should be included. The depth of a submerged cache ordinarily should be specified, but the depth of a buried cache should not be included unless it is buried deeper that the usual 45 centimeters.

Commo cache in three holes in "Y" province "X"country in cemetery three kilometers east city"A" on north side route five. Cache is in northeast corner near walled plot. Container one is west of plot one two meters from northwest corner and one six meters from southwest corner. Container two is four meters west of southwest corner in line with south side. Container three is one south side adjacent to southeast corner of plot.

Figure D-3. Sample message data.

APPENDIX E

SECURITY SURVEY OUTLINE FORMAT

The sequence of the following outline ensures all areas are covered. However, the *General* section should always be first.

GENERAL

Purpose of the survey and the date–time group._____
_____.

Name of the plan, organization, or entity._____
_____.

Location (*jurisdiction, city, county, federal; street address, mailing address*)._____
_____.

Coordinates.

   • Universal transverse mercator (*grid*)._____
   _____.

   • Geographic._____
   _____.

   • Lambert (*if used, must state*)._____
   _____.

Map references_____
_____.

Security responsibility (*private or governmental agency*)._____
_____.

Type of business and product or services rendered._____
_____.

Brief history of locality._____
_____.

Size and physical characteristics of the locality and the labor force._____
_____.

Previous surveys (*including the date and conducting agency, and a statement of the corrective actions taken*)._____
_____.

E-1

SOCIOPOLITICAL CONSIDERATIONS

Radical or reactionary groups._____

_____.

Local power structures._____

_____.

Primary political groups and opinions._____

_____.

Economic factors of the area._____

_____.

PHYSICAL SECURITY

Terrain and vegetation._____

_____.

Avenue of approach._____

_____.

Perimeter barriers._____

_____.

- Natural barriers (*Location, description, dimensions*)._____

_____.

- Structural barriers *(Location, description, dimensions [length, width, height, thickness], construction material, maintenance conditions).*_____

_____.

- Human barriers *(Type, number, duties, type of defensive equipment, shifts communications [primary, alternate], security training type and frequency).* ___

_____.

- Animal barriers *(if used)*._____

_____.

- Energy barriers. *(Include all sensors, alarms, and monitoring systems, with a separate paragraph for each system, including model numbers and brand names, location, area monitored, overriding areas [if any]), and any areas not covered [dead space]).*_____

_____.

Emergency and evacuation plans and alarm procedures *(fire, bomb threat, rocket attack, mob attack)*._____

_____.

Entrances and emergency exits. *(List all at every floor level [ground level first, roof last])*.___

_____.

- Vehicle entrances. (*Describe in detail*)._____

_____.

- Personnel entrances. *(Describe in detail)*._____

_____.

- Entrance points. *(Describe in detail type of door; direction of opening; dimensions; users; hours of operation; type of frame; hinge location, dimension and types; two—way viewers; locking hardware).*_____
_____
_____
_____.

Interior description and floor layout *(Ground level to roof, staircases, elevators, and corridors).*_____
_____.

Windows *(Type, dimensions, and construction materials; types of lock).*_____
_____.

Interior barriers *(Type, location, areas that are denied access to, when locked, type of door or gate, dimensions, construction material, locking hardware).* _____
_____
_____.

Protective lighting *(Type, adequacy, location, location of control switches, personnel who operate controls, alternate system).*_____
_____
_____.

Water *(Primary and alternate sources, reserves, monitoring procedures, security).*_____
_____.

Power *(Primary and alternate sources, location and access to sources and main distributing points; personnel who operate and maintain system; emergency generator model, type, fuel used, and fuel consumption; location of fuel storage, security of storage, capacity).*_____
_____
_____.

Food sources and reserves *(Primary and alternate stockage, types of food, resupply schedule and procedures).*_____
_____.

Communications *(Telephone, very high frequency (VHF), high frequency (HF), messengers).*_____
_____.

Fire fighting facilities *(Personnel, equipment, alarms, external assistance).*_____
_____.

Air conditioning and ventilation._____
_____.

PERSONNEL SECURITY

Key personnel._____
_____.

Security clearances._____
_____.

Morale *(employees)*._____
_____.

INFORMATION SECURITY

Classified information._____
_____.

Areas where important documents are kept *(controlled access)*._____
_____.

Safes, cash registers._____
_____.

Transmission of documents from one location to another._____
_____.

SECURITY EDUCATION

Amount of security education given._____
_____.

Persons receiving security education._____
_____.

Frequency given._____
_____.

Document destruction procedures._____
_____.

SAFETY

Safety inspections._____
_____.

Safety reports._____
_____.

Safety organization and responsibilities._____
_____.

MEDICAL

First aid station._____
_____.

Nearby hospitals._____
_____.

ANALYSIS, FINDINGS, AND RECOMMENDATIONS

BLOCK FACE AND NEIGHBORHOOD
CHARACTERISTICS SURVEY
OUTLINE FORMAT

GENERAL

Description of the area.

- General description of the area and location within the city._____
  _____.

- Terrain, elevation, vegetation, and characteristics *(urban, suburban, or rural)._*
  _____.

- Type of neighborhood *(business, residential, industrial, institutional, recreational, underdeveloped, ordeteriorated).*_____
  _____.

- Boundaries of the survey area._____
  _____.

- Location of the principal structure being surveyed._____
  _____.

Map references *(United States geological survey maps, local maps).*_____
_____.

Name of surveyor and date. *(describe who conducted the survey and date of survey).*_____
_____.

Additional information.

- Population mix and distribution *(dress code, age group, income level, crime [high, moderate, low, organized, or random]).*_____
  _____.

- Climate *(temperature range and mean, relative humidity, rainfall, snow, onset and end of freezing, floods, and tornadoes).* _____
  _____.

- Political attitudes, riots, strikes, mobs *(effectiveness and ability of local services; frequency and thoroughness of patrolling activity and response time).*_____
  _____.

- Transportation systems and facilities *(location of bus stops, bus schedules, taxis, or other).* _____
  _____.

- Block face characteristics *(lighting, level of traffic [vehicular and pedestrian], parking areas [above ground and underground], kinds of buildings, alleyways, recreation facilities, education).*_____
  _____.

F-1

221

SPECIFIC AREA INFORMATION

Potential for natural surveillance.

- List and describe each location *(type of building, size, distance, and direction from principal structure)*._____
_____.

- Describe what can be observed and what the advantages and disadvantages are *(water, lights, rest and recuperation areas, cover, concealment, concealed entrances)*._____
_____.

- Consider whether a sniper will have to contend with plunging fires, or are they advantageous to him._____
_____.

- Depict each position in the sketch and take photographs._____
_____.

- Obtain a point of contact._____
_____.

- Describe how to gain access._____
_____.

Potential to house an assaulting force.

- Determine what size of force can be staged at the location and for how long.____
_____.

- List and describe each location, applying the same considerations as in the *general* description._____
_____.

- Depict in photos and sketches._____
_____.

Avenues of approach and evasion.

- Describe any covered and concealed routes to and from potential surveillance and staging location to the principal structure._____
_____.

- State which route offers the best protection._____
_____.

Safe sites.

- Determine if there is a safe location nearby to evacuate to in the event of mob assault._____
_____.

- List and describe each location *(type of building, distance from principal structure, routes to and from, how to gain access, point of contact if possible)*._____
_____.

• Depict safe sites in photos and sketches._____
_____.

Natural and man-made obstacles and potential danger areas._____
_____.

Additional information._____
_____
_____
_____.

GLOSSARY

Section I. Acronyms and Abbreviations

AO	area of operations
AP	antipersonnel
AR	Army regulation
AT	antitank
CARVER	criticality, accessibility, vulnerability, and recuperability
CEOI	communications–electronics operation instructions
CEWI	combat electronic warfare and intelligence
CID	Criminal Investigation Division
CMT	crisis management team
CO	commanding officer
CP	command post
CS	ortho–chlorobenzalmalononitrile
DA	Department of the Army
DZ	drop zone
EOC	emergency operations center
EW	electronic warfare
F	Fahrenheit
FAC	forward air controller
FBI	Federal Bureau of Investigation
FEBA	forward edge of the battle area
FID	foreign internal defense
FM	field manual
FRP	final reference point
GSR	ground surveillance radar
HF	high frequency
HQ	headquarters
IRP	initial rallying point
IR	infrared

IVDS	Installation Vulnerability Determining System
LP	listening post
LZ	landing zone
MACOM	major Army command
METT-T	mission, enemy, terrain, troops, and time available
MLD	main line of defense
MPI	military police investigator
MSS	mission support site
NBC	nuclear, biological, and chemical
NOD	night observation device
OCOKA	observation and fields of fire, cover and concealment, obstacles, key terrain, avenues of approach
OP	observation post
OPORD	operation order
OPSEC	operations security
ORP	objective rallying point
OT	observer target
PSYOP	psychological operations
PW	prisoner of war
R & S	reconnaissance and security
RATELO	radiotelephone operator
RDF	radio direction finding
recon	reconnaissance
REMS	remotely employed sensor
RON	remain overnight
RP	rallying point
RRP	reentry rallying point
RSTA	reconnaissance, surveillance, and target acquisition
RT	radiotelephone
RZ	reconnaissance zone
SALUTE	size, activity, location, unit or uniform, time, equipment
SERE	survival, evasion, resistance, and escape
SF	Special Forces
SFOD	Special Forces operational detachment
SH	student handbook
SIR	specific information requirements

Glossary-2

SITREP	situation report
SJA	Staff Judge Advocate
SO	special operations
SOP	standing operating procedure
SRT	special reaction team

TACAIR	tactical air
TC	training circular
TM	team
TMF	threat management force

UV	ultraviolet
UWOA	unconventional warfare operational area
UW	unconventional warfare

| VHF | very high frequency |
| VR | visual reconnaissance |

| WP | white phosphorous |

Section II. Definitions

Active and passive. RSTA equipment is also subcategorized as either active or passive.

- *Active RSTA equipment* projects energy to detect a target. This energy can normally be detected by an enemy.

- *Passive RSTA equipment* either detects energy being projected or uses available energy as a detection means. This equipment is not detectable by an enemy.

Ambush force. The patrol, squad, platoon, or other unit that establishes an ambush.

Ambush of opportunity. The ambush of a target of opportunity, often the action of a search and attack patrol.

- When available information does not permit the detailed planning required for deliberate ambush, an ambush of opportunity is planned. In this case, the ambush patrol plans and prepares for the ambush and attacks the first suitable target appearing.

- Before departing, a search and attack patrol plans and rehearses the ambush of the types of targets it may encounter. It establishes and executes ambushes as opportunities arise.

Ambush site. The terrain on which a point ambush is established.

Antiterrorism. The preventive measures taken to reduce the probability of a terrorist act occurring.

Attack force. The fire and maneuver portion of a point ambush. In a patrol, the assault and support elements are the attack force.

Countering terrorism. Systematic measures taken to reduce terrorist incidents from occurring on military installations.

Crisis management team. A team at major commands or at various command installation levels concerned with plans, policies, procedures, techniques, and controls for dealing with sudden violent acts of terrorism on istallations and facilities. The team considers the local, national, and international implications of major disruptions, and establishes contact with the Army Operations Center as the situation escalates and requires higher level involvement and guidance. Normally at installation level the CMT is established at the designated emergency operations center.

Deliberate ambush. An ambush planned as a specific action against a specific target. Detailed information of the target is required: size, nature, organization, armament, equipment, route of movement, and times the target will reach or pass certain points on its route. Deliberate ambushes are planned when—

- Reliable information is received on the intended movement of a specific force.

- Patrols, convoys, carrying parties, or similar forces establish patterns or size, time, and movement sufficient to permit detailed planning for their ambush.

Destruction ambush. An ambush that includes assault to close with and decisively engage the target.

Far ambush. A point ambush whose assault element is located beyond reasonable assaulting distance of the killing zone (beyond 50 meters is a guide figure). This location may be appropriate in open terrain offering good fields of fire or when attack is by fire only (harassing ambush).

Harassing ambush. An ambush in which the attack is by fire only.

Killing zone. The portion of an ambush site where fires are concentrated to isolate, trap, and destroy the target.

Near ambush. A point ambush whose assault element is located within reasonable assaulting distance of the killing zone (50 meters is a guide figure). Close terrain, such as jungle and heavy woods, may require this positioning. It may also be appropriate in open terrain in a *rise from the ground* ambush.

Negotiations. A discussion between authorities and a (barricaded) offender or terrorist leading to an agreement concerning the release of hostages and the surrender of the offender.

Night observation. Night observation, as it applies to RSTA, means systems (night observation devices) that aid visual observation during the hours of darkness. These devices can be used in the surveillance or target acquisition role.

Offender. Any person who commits a special threat act; this includes snipers, barricaded lawbreakers or terrorists, and hostage-takers.

Rise from the ground ambush. A point ambush in which the attack element is completely concealed in the *spider hole* type of covered foxholes. When the ambush starts, the assault element throws back the covers and *rises from the ground* to attack the target. This ambush is appropriate in open terrain, seemingly void of good cover and concealment, and otherwise lacking the features considered desirable in a good ambush site.

Security element. The early warning and security portion of a point ambush. In a patrol, the security element is the security force.

Special reaction team. Any team of military or security personnel specially trained, armed, and equipped to contain and neutralize a special threat. Normally a team of military police personnel accomplished in the use of special weapons and tactics. The SRT is part of the inner-perimeter security team and performs the assault when required.

Special threat. Any situation involving a sniper, barricaded lawbreaker or terrorist, or hostage-taker(s) that requires special reaction or response, manpower, and training.

Surveillance. Surveillance is the continuous all-weather day and night watch over the battle area. Personnel engaged in a surveillance mission usually report the information to the S2 or G2 at battalion, brigade, or division. The information they collect contributes to an overall knowledge of the enemy situation.

Target acquisition. Target acquisition is that part of combat intelligence that pertains to detection, identification, and location of a target in sufficient detail to permit the effective employment of organic and supporting weapons.

Threat committee. An important instructional element to threat analysis useful in centralizing efforts to harden or eliminate discovered weaknesses. These efforts may include brainstorming, or a think-tank approach to analyzing the threat. The committee can be made up of anyone possessing skills and knowledge in combatting terrorism.

Threat management force. An action force drawn from installation resources that responds to major disruptions on the installation. The TMF is commanded by the on-site commander who may be the installation provost marshal. In the area of counterterrorism, the force includes the initial response, innerperimeter security element, outer-perimeter security element, special reaction team, and negotiations team.

REFERENCES

FIELD MANUAL (FM)

5-36	Route Reconnaissance and Classification
7-85	Ranger Unit Operations
19-10	Military Police Operations
19-15	Civil Disturbances
19-30	Physical Security
21-20	Physical Fitness Training
21-60	Visual Signals
21-75	Combat Skills of the Soldier
90-10 (HTF)	Military Operations on Urbanized Terrain (MOUT) (How to Fight)
90-10-1 (HTF)	An Infantryman's Guide to Urban Combat (How to Fight)

ARMY REGULATION (AR)

50-5	Nuclear and Chemical Weapons and Materiel – Nuclear Surety
190-52	Countering Terrorism and Other Major Disruptions on Military Installations
190-53	Interception of Wire and Oral Communications for Law Enforcement Purposes
380-13	Acquisition and Storage of Information Concerning Nonaffiliated Persons and Organizations

TRAINING CIRCULAR (TC)

19-16	Countering Terrorism on US Army Installations
23-14	Sniper Training and Employment
34-50	Reconnaissance and Surveillance Handbook
90-6-1	Military Mountaineering

SPECIAL TEXT (ST)

31–195   Special Forces Reconnaissance Handbook

Source: Commander
         USAJFKSWCS
         ATTN: ATSU–DT–PDA
         Fort Bragg, NC 28307–5000

OTHER

Ranger Handbook

Source: Commandant
         US Army Infantry School
         ATTN: Ranger Department
         Fort Benning, GA 31905

STUDENT HANDBOOK (SH)

7–285   Reconnaissance Surveillance, and
         Target Acquisition Handbook

Source: Commandant
         US Army Infantry School
         ATTN: CATD
         Fort Benning, GA 31905

INDEX

By Order of the Secretary of the Army:

CARL E. VUONO
General, United States Army
Chief of Staff

Official:

WILLIAM J. MEEHAN II
Brigadier General, United States Army
The Adjutant General

DISTRIBUTION:

Active Army, USAR, and ARNG: To be distributed in accordance with DA Form 12-11A, requirements for Special Forces Operations (Qty rqr block no. 531).

☆ U.S. GOVERNMENT PRINTING OFFICE : 1991 - 281-486 (42138)

www.ingramcontent.com/pod-product-compliance
Lightning Source LLC
Chambersburg PA
CBHW071712170526
45165CB00005B/1991

* 9 7 8 1 4 8 1 8 4 6 5 1 6 *